IN MEMORY OF
DR. NATHAN COHEN

Teacher, Scientist, Confidant, Friend

CONNECTING WITH NATURE

A NATURALIST'S PERSPECTIVE

BY ROBERT C. STEBBINS

*Emeritus Professor of Zoology and Emeritus Curator of Herpetology
Museum of Vertebrate Zoology, University of California, Berkeley*

First Edition, October 2009

© 2009 Robert Stebbins
Illustrations by Robert Stebbins and
Anna-rose Stebbins unless specified.

All rights reserved. No part of this publication
may be reproduced or transmitted in any form or by
any means electronic or mechanical, including photocopy,
recording, or any information storage and retrieval system,
without permission in writing from both the
copyright owner and the publisher.

Requests for permission to make copies of any part of this
work should be mailed to Permissions Department,
Llumina Press, 7915 W. McNab Rd., Tamarac, FL 33321

ISBN: 978-1-60594-415-9 (PB)
978-1-60594-416-6 (Ebook)

Printed in the United States of America by Llumina Press

Library of Congress Control Number:2009910268

Book design by Kat Thomas

Nature's Gift to Children

Every child should have mud pies, grasshoppers, water bugs, tadpoles, frogs, mud turtles, elderberries, acorns, chestnuts, trees to climb, brooks to wade in, water lilies, wood chucks, bats, bees, butterflies, various animals to pet, hay fields, pine cones, rocks to roll, sand, snakes, huckleberries, and hornets, and any child who has been deprived of these has been deprived of the best part of his education.

—Luther Burbank

TABLE OF CONTENTS

ACKNOWLEDGEMENTS   XI

INTRODUCTION

    Ecology—A Pathway to Connecting with Nature   1

PART ONE

Connecting with Nature, an Ecological Approach—Goals. Activities. Field Trips. Nature Stories. Nature's Driving Force for Change.

Chapter 1: Early Memories and the Nature Connection   5

Chapter 2: Goals of an Ecological Approach   11

Chapter 3: Developing Awareness and Exciting Interest   17

    A Quiet Time With Nature   21

    Making Observations   27

    Note Taking and Writing Skills   29

    Developing Accuracy in Observation and Description   35

    Schoolyard Nature   42

    Getting Acquainted with the School's Biota   44

       Map-making   44

       Data Gathering   45

       Creating a Food Web   46

       Species Identification Files   46

    Creating a Nature Area   49

    Live Animals in the Classroom   52

       Berlese Funnel   53

       Bush-shaking   54

       World of Small Living Things   56

    Frogs and Their Life Stages   57

TABLE OF CONTENTS

Marking Wildlife for Field Recognition	60
A Spider Story	60
An Ant Study	62
Gardens for Learning and Food	64
Learning from the Past—An Historical Background and Beyond	67
Gardens for Nourishing the Mind as Well as the Body	71
Untended "Gardens" and the Teaching of Ecology	74

Chapter 4: Learning from Successful Cooperative Models 79

The Barstow School District Nature Program	81

Chapter 5: Listening in on a Naturalist's Experiences: Nature Walks. Interacting with Plants and Animals in the Field. Nature Stories 87

Nature Walks and Activities	89
Woodcraft Ranger Guide Conference	90
Combining Works of Nature and Man	99
Humans and the Importance of Soil	102
Field Trips of Goodwill	110
Interacting with Animals in the Field	114
Imitating Animal Sounds	114
Owl-calling	114
Playing Parent to Baby Horned Larks	121
An Amphibian Response	122
Use of Distress Sounds	124
Locating Animals by Triangulation	125
Tracking and Animal Sign	127
Desert Tortoise Stories	131
Searching for Eyeshines	135
Collecting with a Camera	135

Photographing Birds	136
Getting Close to Lizards	137
Cohen's Law	140
A Method for Quieting Lizards	146
For the Insect Enthusiast	148

Chapter 6: Evolution, Nature's Driving Force for Change: How Organisms Adapt to Their Environments   151

Summary of Factors Involved in Natural Selection	154
Learning about Exponential Growth (The Biokrene)	155
Natural Selection: A Graphic Demonstration of Principles	159
Real World Examples of Natural Selection— Industrial Melanism	163
Ring Species—A Snapshot of Evolution in Progress	164
Perhaps an Answer for those Who Question the Existence of Missing Links	168

PART TWO
Nature Bonding: Impediments and Hopeful Prospects.

Chapter 7: The High Cost of Ecological Illiteracy: Isolation of Ecology. Attitudes Toward Nature. Population Explosion.   171

Attitudes Towards Nature	174
The Population Explosion	177
Humans and the Biokrene	178

Chapter 8: Hopeful Prospects: an Historic Message that Almost Succeeded. Some Reasons for Hope. A Land Ethic.   181

Some Reasons for Hope	185

TABLE OF CONTENTS

PART THREE
Suggestions for Some Educational Priorities.

Chapter 9: Educational Responses — **191**
    Establishing and Maintaining a
    Nature Centered Educational Program — 194

Chapter 10: Conclusion — **201**
    To Mothers Everywhere — 206–207

APPENDIX — 209
    East Bay Municipal Utility District Biological Survey — 210
    East Bay Park's Naturalist Program — 213
    Orinda School District and Wagner Ranch Nature Area — 217
    Mojave Max, the Spokes-Tortoise for the Clark County
    Desert Conservation Program — 220
    Dr. Dirt's "Life Lab"—An Environmental Education
    Program with Emphasis on Soil Food Webs — 225

NOTES — 233
    Part 1 — 234
    Part 2 — 239
    Part 3 — 242

BIBLIOGRAPHY — 245

NATURE-BONDING ACTIVITIES

    Activities for Increasing Nature Awareness — 24
    Rediscovering the Leaf — 36
    Getting Acquainted with the Schools Biota — 44
        Map Making — 44
        Creating a Food Web — 46

Creating a Nature Area	49
Live Animals in the Classroom	52
Berlese Funnel	53
Bush-Shaking	54
Studying Frogs and Tadpoles	58
Antlion Pitfall Trap	93
Clock Weeds in Action—Experimenting with Clockweed	97
Calling Owls—Imitating a Great Horned Owl	119
Photographing Birds	136
Lizard Noose Construction and Use	142
Photographing the Captive	145
A Hands-on Reality Check—Wild Oats and Dandelions	156, 157
Selection for Concealing Color	160

SUMMARY OF NATURE ACTIVITIES

Chapter 3: Developing Awareness and Exciting Interest

A quiet time with nature	21
Note taking and writing skills	29
A case for nature drawing	32
Developing accuracy in observation and description	35
Map-making for schoolyard studies	44
Creating a food web	46
Creating a nature area	49
Board covers for attracting wildlife	50
Berlese funnel	53
Bush-shaking	54
Heart rate experiments with embryo froglets	59
Marking wildlife for field recognition	60

TABLE OF CONTENTS

Chapter 5: Listening in on a Naturalist's Experiences

Antlion Pitfall Trap	93
Experimenting with clockweed	97
Plot studies	99
Quartz rocks and cyanobacteria (footnote)	107
Calling owls	114
Hide and seek with owl calls	119
Playing parent to baby Horned Larks	121
Making distress sounds	124
Triangulation to locate animals	125
Tracking animals	127
Searching for eyeshines	135
Photographing birds	136
Getting close to lizards	137
Noosing lizards	142
Photographing lizards	145
Tonic immobility to quiet lizards	146
Catching insects	148

Chapter 6: Evolution – Nature's Driving Force for Change

Reproductive potential using wild oats and dandelions	156–158
Colored chips and natural selection	160

Acknowledgements

Many people have contributed importantly to this book. First, I wish to thank Dr. Alan Harnish, cardiologist, who read an early draft and strongly encouraged me to proceed. I also had guidance from my daughter Mary, who taught elementary school in Lavington, B.C. and my daughter Melinda, who, for many years, has taught secondary school in Adelaide, Australia. She wrote "Developing accuracy in observation and description" (p.35) that appears in Chapter 3. Both daughters also helped greatly in organizing and clarifying the text throughout.

Of special importance is the contributions of persons who wrote, or provided close guidance, on major sections of the book that appear in Chapter 4 and the Appendix, "Learning From Successful Models" (p.81 and 209). These case histories, samples of what can be done to promote nature education, provide important support for the goals expressed. They are Leon Hunter, Steve Abbors, Margaret Kelley, Toris Jeager, Ron Marlow and Dale Sanders (professional titles appear in text). Here, I also wish to thank Kay Yanev, one of my former students, for putting me in touch with naturalist Toris Jeager and the outstanding Orinda Elementary School nature and science programs.

In the section on Field Trips, I acknowledge the contribution of Joanne Taylor and Dr. Mary Bowerman (Combining works of nature and man, p.99) and the written and illustrated contributions of Dr. Ted Papenfuss of U.C. Berkeley's Museum of Vertebrate Zoology (Field trips of good will, p.110), a naturalist with a bent for spreading friendship across the planet.

Special thanks go to Mr. Steve Abbors, former East Bay Municipal Utility District Manager, for providing guidance and insight in the development of the EBMUD's biological survey program (p.210), and for his extensive help, and members of his family Carlene and daughters Alison and Rose, in getting the book's text and illustrations camera and computer-ready for publication. Special thanks

ACKNOWLEDGEMENTS

also goes to Malcom Margolin of Heyday books for his support and guidance on publication— and to Kira Od, a dear friend, for putting me in touch with Kat Thomas, a skilled book designer.

Most of all is the debt I owe my parents. They were a remarkable pair. Dad (Cyril Stebbins, of English background), the scientist, agriculturalist and teacher, emphasized self-discipline, civic values ("The Pillars of Society are the home, the school, and the church"), and respect for others and nature. He believed in, and taught, the evolutionary origin of humans. He also sang in the church choir and fully supported the interdenominational church in our community, becoming a favorite of the pastor.

Mom, a Swiss immigrant (Louise Beck), was our loving caretaker, devoted to her brood (We grew up seven—3 boys, 4 girls, in that order, I the oldest). She taught us, humility, caring for others, and "do onto others as you would have them do onto you." She loved wild nature and believed it was God's handiwork. My parents never disagreed on human origin. Evolution was evidently God's way of creation.

Mom loved to tease us—one of her ways of showing affection. I was at her bedside at her death. She was 87 and dying of many small strokes – lucid at times then lapsing into a coma. On her last episode of consciousness she asked me to play my violin. I'm not that good, but I'm sure she wanted to hear it. I said, "I'm sorry Momma I didn't bring my violin. I'll whistle you a little tune," whereupon I whistled "Yankee Doodle." I have no idea why! When I stopped, there was a long silence. I thought she had left us. Then she said, always blunt and to the point, "I like the birds better." I took it as a last affectionate good-bye.

This book and author had the indispensable help (both secretarial and intellectual) of five outstanding assistants: Daniel Mulcahy, Joyce Bautista, Raul Edwardo Diaz, Jr., Keishia Sheffield, and Debbie Lee, mostly supported by U.C. Berkeley Faculty Research Grants. My former graduate students Charles Brown and Sam McGinnis gave me professional help and support.

The butterfly images on pages 209 and 233 are the needlepoint work of my wife, Anna-rose.

INTRODUCTION
Ecology—A Pathway to Connecting with Nature

The Importance of Ecological Literacy and the Whole Organism Approach to Biology

An ecological (whole organism) approach to biology tells us that, whatever else we may be, we are subject to the same laws of nature as are all other organisms; that we are related to everything that lives and are but a single, albeit remarkable, strand in the complex web of life; that our fate is inseparably tied to that of the earth's living community; and that human attributes and behavior have ancient roots extending back to the beginning of life on earth. Every person alive today is a unique manifestation of this remarkable unbroken chain. A nature-centered world view contains the knowledge and ways of thinking that can help us deal objectively, yet compassionately, with our manifold social and environmental problems. Its ethics include and extend beyond our own species to the other life of the planet.

Ecology is the branch of biology that deals with the relationship of living things to their environment and to each other. It focuses on whole organisms (plants and animals) and their interaction in nature. It, and related disciplines such as natural history, botany and zoology, deal with biological subject matter with which people can make direct contact and form emotional ties. This can lead to respect for nature and planetary stewardship.

Unfortunately these participating disciplines, and especially natural history, have declined in popularity at many colleges and universities. They have been considered old-fashioned. Now the whole organism (organismal) approach to biology and its important role in connecting us to nature is needed more than ever before!

In many ways we may be drifting away from our biological roots and in the process, because of our numbers and demands, threatening ourselves and other life of the planet. *The whole organism decline in teaching thus carries with it high physical and emotional costs. Prompt remedial action is called for.*

Richard Louv, in his outstanding recent book, "Last Child in the Woods" extensively documents the importance of nature and outdoor play for the healthy development of children—to prevent what he calls "nature deficit disorder," a malady associated with loss of connection with the natural world. However, he also tells of the efforts of the many individuals and groups working throughout the U.S. and abroad to establish nature in urban and suburban locations—and both the challenges and many successes they have experienced. The whole organism ecological approach to nature education can participate in, and greatly assist, such efforts.

In what follows I offer activities and examples of things I have found effective in getting people interested in nature. Most can be applied wherever there is incentive to do so (See p. XI–XII). I also discuss some major impediments to nature-bonding that threaten our future world-wide—nature-dominating attitudes or indifference toward nature and our failure, so far, to adequately control our population growth. Some hopeful prospects for change are described, but far greater efforts are required, and urgently! (See p. 178–180)

Finally some educational recommendations are offered: (1) Interdisciplinary cooperation by higher education to integrate fields of knowledge so that they are not taught completely in isolation (See p.193). (2) Devising and experimenting with methods to foster ecological literacy, environmental concerns, and nature bonding (See p.197). (3) Finding ways to energize the educational community at all levels (beginning at home) to achieve these goals.

PART ONE

AN ECOLOGICAL APPROACH

GOALS. ACTIVITIES.
FIELD TRIPS.
NATURE STORIES.
NATURE'S DRIVING
FORCE FOR CHANGE.

EARLY MEMORIES AND THE NATURE CONNECTION

My own journey to ecological literacy began in early childhood and continues to this day as a major focus of my life.

CHAPTER ONE

EARLY MEMORIES AND THE NATURE CONNECTION

One of my earliest memories is of a warm day, a field with many grasshoppers, a shallow creek with cold water, and the joy of a day in the hills with my parents. My dad had gone fishing, and I was free to wander about nearby. It was summer in the Gray Pines foothills of the Sierra Nevada, near Chico, California, where I was born (March 31, 1915). Along the creek I found a turtle! I had hoped someday to have one as a pet. I ran with the wondrous creature cradled in my hands to show my Mom. I was enthralled with its bright eyes, the feel of its claws, and its cold body as it struggled to free itself from my grasp. So began a lifetime of connecting with nature.

On our Chico Ranch, at age five or six, my dad taught me the parts of a flower, an almond blossom - a first sex lesson about plants. He taught me the names of the birds that frequented our fields and orchards, and their importance to our crops in control-

"Nature Boy" with pets.
Sherman Oaks "Ranch," 1927

ling pests. I got my first spanking when I left the care of my baby sister to follow a Great Horned Owl into our almond orchard. When our barn burned down came a lesson on spontaneous combustion. Dad thrust my hand into a pile of green alfalfa to feel the heat within. Hired help had put green cuttings in the hayloft!

With younger brother Hubie, we made tunnels and played hide and seek in our patch of alfalfa used to feed the livestock. The mower cannot pick up flattened alfalfa. So disregard of parental warnings led to another spanking. When our growing family, now with six

EARLY MEMORIES AND THE NATURE CONNECTION

siblings, was ready to put down roots in Sherman Oaks, in the foothills of the Santa Monica Mountains overlooking the San Fernando Valley, I insisted (at age 10!) there must be a big tree on the property. A giant wild Black Walnut with a trunk over five feet in diameter, the largest of its kind I ever found in the hills was ours! Like young monkeys, we boys (now three) and usually several neighbor kids, throwing caution to the wind, played risky, but exciting tree tag in the big tree. Soon a tree "house" was built, anchored high in the strong upper branches—a platform with fenced sides, open to the sky. On warm nights we slept on a hay mattress under the stars. It was here I learned the song of the Western Screech Owl, and was able to attract the bird with my imitations.

When I was 12 I felt like a wild thing, and sometimes ran through the hills barefoot (my soles like leather) and in my swimming trunks, even in the rain. I felt like I would never get tired. A favorite time for me was in spring when the air was scented with the flowers of the chaparral and the birds were nesting. I loved to sleep out in the hills then, to hear the bird dawn chorus—nature's orchestra that plays at dawn in many parts of the world.

It begins before sunrise, so I tried to be awake to hear the first note, soon to be followed by many others in a growing crescendo with the arrival of daylight. Knowing most of the songsters added greatly to the enjoyment of this wonderful sound fest and I always felt that I had entered nature's cathedral, even though as a budding naturalist I knew I was hearing a glorious declaration of territorial rights.

We had a menagerie of animals on home ground—goats, rabbits, pigeons, turkeys, chickens, ducks, white rats (housed in an apartment of my making), cats and always a dog. I raised several baby owls brought in from the hills. Catering to my growing naturalist interests my Dad hauled in an old chicken roost, that with some reconstruction turned into a museum—housing freshly dead animals I stuffed, abandoned bird nests and addled eggs I collected and emptied of contents, mounted butterflies and other insects caught, wild flowers pressed, and other treasures dear to the heart of a young boy turned loose in wild nature.

EARLY MEMORIES AND THE NATURE CONNECTION

Our "ranch" also contained a variety of plants that helped feed the growing family—potato and tomato patches, fruit trees (providing peaches, nectarines, oranges and lemons), grapes, black berries, strawberries, and squash and watermelons (so many we sold to neighbors). Dad, the agriculturist, science teacher, and nature-study expert, had chosen our land wisely—a soil rich alluvial fan at the mouth of Madelia Canyon, its richness indicated by the great Black Walnut.

We children, now six, despite our fortunate circumstance, learned early on that nature and the human condition are not all goodness and light. As much as we loved being in the hills, we soon learned that there were rattlesnakes to look out for, mountain lions were still present, that we should not tamper with bees and wasps nests, and that poison oak was a plant to be avoided. (On a dare sister Lola, thinking she was immune, rubbed her face with its leaves and broke out with a terrible rash). I brought the "seven-year itch," an infection caused by a skin burrowing parasite, to all siblings, presumably from sleeping, mostly bare, on an old mattress in an abandoned disintegrating cabin in the hills. There followed a week of rubbing us all down, naked, with a sulfur/lard compound and washing all clothing and bed sheets everyday. What a task for Mom!

A cloud burst in the headwaters of the canyon above our home caused a flood that scoured out all vegetation to a height of 8 to 10ft., took out most of the small animal life and dumped branches, rocks, and soil too close to home for comfort, reminding us how our soil-rich alluvial fan had formed.

Thus connecting with nature entails not only admiration of the beauty and complexity of nature, but the need for understanding and respect. In its broadest sense "nature" embraces everything from atom to cosmos and life in all its manifestations including us and our creations. However, <u>as I use the term throughout this book my focus is on the planet's living system—an ecological approach</u>. It's life itself that commands our greatest attention.

The Biophilia Hypothesis

For thousands of years we, the "knowledgeable" animal (*Homo sapiens*), the toolmakers, were nature's children. In small bands we hunted and gathered our food and lived at nature's sufferance. Life was often difficult and fraught with danger, but earth's wild plants and animals fed, sheltered, and clothed us and we learned to avoid those that caused us harm. Wild plants were our medicines. <u>Our lives absolutely depended on our knowledge of wildlife (plants and animals), thus we became the first ecologists!</u> When we increased beyond nature's support, we learned to control our numbers, fought our rivals, or moved elsewhere. Over the long span of our development such survival demands must have fine-tuned the circuitry of our brains and given us a deep affinity toward the wildlife that sustained us. That affinity scientists now call *"biophilia"** (an affiliation toward life). Its presence among our early forbearers was expressed in totems, creation stories, rituals, and pictographs, and in their homage for the animals they killed. In those days our spiritual connection with the Earth and its life ran deep.

Fast forward to the present. We, the "knowledgeable" tool makers, in a flash of time have overrun the Earth and the separation emotionally of many of us from wild nature, especially in the "developed" world, does not bode well for our future. Yet deep inside, even in the most dedicated urbanite, must reside that longing for nature (biophilia), the legacy from our ancient past. Our task now is to reduce nature alienation gaps and to create an educational process that will help us get there.

---

*Some scientists believe humans may harbor an innate affinity for other species—what biologist E.O. Wilson calls "the urge to affiliate with other forms of life," or "biophilia." (See Wilson, 1984)

GOALS OF AN ECOLOGICAL APPROACH

Dealing with nature's complexities

CHAPTER TWO

GOALS OF A ECOLOGICAL APPROACH

Living systems are complex thus tampering with them calls for great care. "The do no harm principle" should be paramount. Through chain reactions and unexpected interactions, widespread and sometimes catastrophic effects can occur (See Ausubel, 2004). Destruction of a species of ant reduced the productivity of European forests*[1] (P.1), and introduction of the European fox had far-reaching impacts on native marsupials in Australia. The spread of non-indigenous plants, animals, and microbes is a rapidly growing problem of global proportions—fueled by climatic change, commerce, and an expanding, and increasingly mobile, human population. It is considered by many biologists to be one of the greatest threats to biodiversity, planetary ecosystems, and public health.

Although the basic principles of ecology upon which nature-centrism is based are readily comprehended (if taught), the organisms themselves and the interrelationships among them and their environments are not easily understood without careful, and sometimes prolonged, field training and study.

Consider the incredible interlocking hierarchy of increasing complexity and interaction of physical and chemical properties and environmental influences in going from DNA, chromosomes, cells, individual organisms, population, community, ecosystem, to the biosphere!

Field studies are absolutely essential to understanding nature and of increasing importance as human impacts escalate. Long-term studies are particularly useful in assessing ecological trends and problems*[2] (P.1). Often combinations of field and laboratory studies are required to obtain satisfactory results.

A supportive public and a political, educational, and scientific community, that appreciates the importance of ecological and environmental knowledge, can bring about the educational and investigative processes, and the training of people, necessary to achieve such long-term goals (See Roving Professional Naturalists, p.198).

---

*See Notes Section, p.233

Conveying Reverence Toward Nature

Nature-centered responses work best if they have emotional roots, and grow from deep-seated feelings based upon understanding, respect, and reverence toward nature.

A powerful pro-environmental coalition may be emerging. Religious people and institutions, scientists, and advocates of sustainable development are beginning to partner. They share an appreciation of nature that surpasses its economic value. A new ethic encompassing humans, the divine, and nature may be in the making. See (3 P.1), Global Forum of Spiritual and Parlimentary Leaders, p.236 and Gardner, 2006, Chapter 5, Nature as sacred ground).

Aldo Leopold warned against an arid conservation, "which... defines no right or wrong, assigns no obligation, calls for no sacrifice, implies no change in the philosophy of values" (See also Gorman, 2003).

Implementing a Nature-Centered Educational Program

The nature-centered educational program is partly an affirmation of the values of "nature study", as pursued in the last century (See Comstock, 1915), but it calls for some important modifications.

A Rebirth for "Nature Study"?

In the 1800s and, to some extent, into the early 1900s, "nature study" dominated elementary school science, but under the momentum of great technological advances, it gave way to a more technologically oriented science program. In reference to this shift, I remember well the remarks of a young science educator who spoke at a school science conference in the early 1960's. He said, "Soon we will be traveling thousands of miles per hour—thank goodness we have gotten away from the grasshoppers and butterflies in elementary school science." If we have indeed left the grasshoppers and butterflies, we must now return to them, but in ways somewhat different from in the past.

GOALS OF A ECOLOGICAL APPROACH

There was much of value in the old nature study program—the direct observation of local plants and animals, their structure and manner of life—but it often lacked the unifying theme of evolution and was sometimes given to moralizing, excessive sentimentality, and interpretations based on the way humans think and behave.

In what ways should nature-centered studies differ from those of the old nature study program? They should emphasize:

1. Ecological principles and the interrelationships of the organisms studied;

2. Involve students and teachers in <u>actual studies</u> of living things and their responses and habits <u>in or near</u> the school or home environment and on trips afield;

3. Focus on learning the methods of science, the "discovery" (lets find out, hands on) approach to learning rather than mainly the "telling" approach; and

4. Give attention to human ecology, humanity's place in the web of life, including the ecological impacts of human actions and attitudes toward nature.

DEVELOPING AWARENESS & EXCITING INTEREST

CHAPTER THREE

DEVELOPING AWARENESS & EXCITING INTEREST

The chapter offers many examples of effective programs and approaches that can help restore and expand ecological connections—among them gardens planted and tended by the students, nature field trips, and other outdoor activities as learning tools. It is not just for teachers or professional educators nor does it offer a formal curriculum. It is for anyone interested in ways to learn more about living things and how such learning can help contribute to a safe and happy future (See p. VIII–X, Index to Nature—bonding activities).

Start Early—Begin with Children

Nature-centered training with emphasis on living systems should begin in childhood, when basic attitudes take root. It should start in infancy by encouraging a young child's interest in living things, and it would be desirable as a common thread from kindergarten through high school. As parents we owe it to our children to encourage such interests. *"Caring for creatures or plants is often the child's first experience in the feelings and practicalities of caring for another, as well as in responsibility."* (See Kahn Jr. and Kellert, 2002, and Louv, 2005, 2006). Children will pay a high price for ignorance about their place in the natural world!

Group outings in natural environments, with its associated comradery at an impressionable age, can have lasting impact. This is why elementary school nature study and environmental field trips can be so important. Schools should not assume that such experiences will be cared for by family or youth group activities alone. Many children are denied such experiences. The elementary grades are especially important because a large segment of the youthful global population has no additional schooling. One of the ways our teacher daughter Mary made sure her students would experience nature first hand was through a yearly camping trip.

Regrettably, field trips seem to be the first to go when budget cuts are required. Impediments are test-driven curricula and the rising cost of insurance and transportation. Given the problems we now face, school districts should seek ways to remedy these problems (For an example, see p.81, The Barstow School District).

DEVELOPING AWARENESS & EXCITING INTEREST

Anyone can begin the steps toward a more nature-centered world view. Start by looking away from time to time from the people-centered world. Look at the stars (hard to do in city environments), look for the small life in the soil in the backyard, search the leaves and stems of plants for insects, examine toadstools in the lawn, and watch and listen to birds. If you are a beginner, get acquainted with some of the non-domesticated plant and animal neighbors and learn their names. You may feel you do not have time for such things, but even small steps can count. I can seldom look at a flower now without recalling my childhood memory of my father teaching me the parts of a flower. Children need support and guidance in connecting with nature (See Carson, 1998).

With very young children nurturing their sense of curiosity and wonder should be a primary goal. Let's not let those early formative years slip by without helping them to see the beauty and remarkable diversity of life around them. At this stage, feeling is more important than knowing. Don't hesitate because you don't know the names of the players*. It is enough to see and appreciate. A hand-lens magnifying glass will help. Try letting the child lead.

"Curiosity is a delicate little plant which aside from stimulation stands mainly in the need of freedom." —Albert Einstein

"The old Lakota was wise, He knew that man's heart away from Nature becomes hard. He knew that lack of respect for growing, living things soon led to lack of respect for humans too. So he kept his youth close to its softening influence."
— *Chief Luther Standing Bear (Lakota)*

---

*Parents and other leaders should at least know the hazardous plants and animals in their surroundings. (Check with authorities.)

A Quiet Time With Nature:

The "Solo" Experience—Alone with Nature

Rediscovery's Solo/Vision Quest

"Vision questing" has been the traditional rite of passage on the North American continent for thousands of years. "The Indians marked the transition from childhood to adulthood with long periods of isolation, fasting, and meditation in the wilderness in search of spirit guardians" (Henley, 1996).

A modern and modified approach to the "rite of passage" are adventure programs of "rediscovery" that emphasize the "solo" experience (the individual, alone and abroad, over night in a wild place), undertaken through careful guidance of native and/or non-native elders or overseers.

The process involves reconnecting with the Earth and humanity—"to discover the world within our self, the cultural worlds between people, and the wonders of the natural world around us" (Henley, 1996). Special emphasis is on youth—native and non-native, but all ages can participate.

The Solo/Vision Quest has become Rediscovery's greatest key for unlocking the doors to self-discovery. A young girl who spent the night curled up under a spruce tree wrote, "I thought the earth remembered me. She took me back so tenderly" and a boy wrote, "Here I am on solo night. At first it gave me quite a fright but when I see the swaying trees and hear the peaceful rhythm of the sea, I feel more confident in me." A Haida elder told a group of soloists, "If you can learn to live in harmony with this experience, it will carry over into the rest of your life."

It is of interest that over a decade of Rediscovery programs showed "that the younger the person, the more likely they are to volunteer for Solo and the more likely to successfully complete it," and that bullies or individuals at the top of the "pecking order" will almost always fail. A native elder explained it this way: "The reason a child acts this way is because he doesn't know or like himself. Nobody wants to spend that much time alone with someone they don't know or like—especially if that person is yourself!"

Finally, a word of caution: Although the Solo experience can be profound in connecting with nature, care must be exercised to insure safety of participants. Sites chosen must avoid natural hazards and knives and hatchets should be forbidden, especially with younger children. A "security" campsite should be nearby for those who wish to come in before morning.

MVZ's Solo Exercise

A "solo" class exercise was conducted in our course on Vertebrate Natural History at U.C. Berkeley, California, to focus the student's senses on nature. Our project developed independently of the Solo/Vision Quest. It took place at daytime when the weather was pleasant and our students had become comfortable with being alone in the field and acting independently, undistracted by other students or faculty. By this time they were aware of plants and animals to be avoided. Field trips were conducted in Tilden Regional Park and in other parks in the East Bay Hills, California.

We expected that under these conditions their curiosity about nature would be further aroused and the experience would help them formulate questions that might be answered through careful field observation. Each student was to select a research project to be reported on at the end of the course. They also were to write a brief report on their "nature awareness" experience.

Solo activities can be done <u>anywhere</u> there are remnants of wild nature and where it is safe to do so. Our students knew the hazards of poison oak and ticks.

To Danny, one of my students, who learned the meaning of "solo" independently, without a teacher. Early in life, he became a forester.

Activities for Increasing Nature Awareness

Step 1. Find a safe private place, preferably in a more or less natural area. Spread out a ground cover upon which to sit or lie, large enough so no part of your body will extend beyond its margins. This helps to avoid ticks and some biting or stinging insects. Sit or lie down on your back with eyes closed. Relax and try to clear your mind of extraneous thoughts. Focus on sound. With eyes closed listen to the general sounds in the environment. Try to distinguish between human and non-human sounds. Note location, frequency, volume, and identity (if possible)—someone chopping wood, a dog barking, bird singing, crickets, sound of wind in trees or grass, etc. Still resting with eyes closed, savor your auditory input for a while. Already you may feel a calming of spirit.

Step 2. Focus on sight and touch. Lying on your stomach, examine the nearby environment just beyond the edge of your ground cover—nearby grass, dead leaves, etc. Watch for small animals—an aphid sucking plant juices, a ladybird beetle climbing a grass stem, an ant making its way through the tiny jungle of grass. After 5 minutes or so, feel the texture of dead leaves, grass stems, pieces of bark and fallen twigs, etc. Try to imagine what this tiny world would look like to an ant. Again keep your mind off other things. The calming effect should continue.

Step 3. Focus on smell. Get up and walk about in the small area of your immediate surroundings. Is there a general odor you can characterize? Sniff and feel the bark of trees, crush some leaves of different plants and note their odor. Look about you. Savor being alive in this beautiful place.

We usually allowed about an hour for this exercise.

DEVELOPING AWARENESS & EXCITING INTEREST

A somewhat similar approach was used by Matt Miller, Director of Communications, The Nature Conservancy, Idaho (See Gorman, 2003, Awakenings. Spiritual perspectives on conservation. Nature Conservancy, Vol. 53, No.1, p.22). He was teaching a nature literature and writing class at a small liberal arts college in Kansas. "I had the students go out on the prairie and sit for two hours without a book, a cell phone, or any activity. Some of them had never done anything like this before, and I got all kinds of complaints. But their essays were some of the most spiritual writings I'd ever read."

A quote from a student report on our "nature awareness" experience in Tilden Park supports Matt Miller's observation.

"The exercise questions do not mean much to me today. But lying on this spot of earth, I am alone and together with everything that is alive here. The sight of waving wild flowers, grasses, and trees; the roar of wind blown trees, the fragile songs of the birds proclaiming their right to existence, the hawk soaring above, the scent of the meadow, the raspy touch of grasses and the solidness of the earth—all stream into my senses. I am asked by each of these—how can I save it? What am I to do with my life to ensure the lives of the meadow? Briefly my thoughts travel to careers—will the one I choose really help save this aliveness? And I am brought back home in the calls—high and melodious—of some bird and I am alone again and together with everything that is alive here"

Implementing a nature-centered program in many places now faces the problem of increasing urbanization. It is predicted that by 2025, around 60 percent of humanity will be living in urban environments (United Nations Estimate). This challenges efforts to promote nature-centered thinking, but we must be successful, especially in cities, because in democratic societies, the fate of the life of the planet is increasingly now dependent on the attitudes of city dwellers.

DEVELOPING AWARENESS & EXCITING INTEREST

Fortunately, there are often surprisingly large numbers of free-living animals (those living on their own) even in cities: insects, spiders, and other invertebrates and vertebrates such as birds*, frogs, mice, and others. Many are associated with native and introduced plants. These animals and plants, and their interrelationships with the physical and living environment, including their interactions with people, can be studied. However, it is easy to overlook what has become commonplace. Before undertaking studies, spend some time looking over the immediate outdoor environment or a nearby area, to see what is available. For safety concerns, check with local wildlife authorities to see if there are plants or animals in the area that should be avoided.

> *Imagine yourself an explorer that has entered an unexplored wild place. Because of the complexity inherent in nature, one's backyard or a schoolyard may be such a place. Begin by looking at common species of plants and animals, including the small world of nature seen through a hand lens, on trips afield and in the home and school environment. One need not have spectacular or exotic subjects. There are many mysteries to be uncovered among the small invertebrates found in one's immediate surroundings.*

With a color scanner and a digital camera you can provide a beautiful ongoing record of plants and animals identified in your yard, on walks in the countryside, and visits to favored vacation sites. Communicate with other folks on the Internet who are also engaged in the activity to share experiences, knowledge, and even photographs. However, many families are so harried now that time may be a deterrent, but start with small steps—one species at a time. Consider the value to children in helping to ensure that they do not lose their connection with nature. Build a long-lasting album of wildlife encounters.

---

*See the video "Pale Male" a Red-tailed Hawk that took up residence in the Central Park area in New York City, and over a period of 13 years attracted a total of four mates and together raised 23 offspring. (Produced by FL Productions © 2004 Thirteen/WNET, New York). Red-tailed hawks have continue to expand in the area.

DEVELOPING AWARENESS & EXCITING INTEREST

Making Observations

There are several things of primary importance in making observations: (1) applying one's senses closely and thoughtfully to what is observed; (2) patience—sustaining observations long enough to see what is actually happening; (3) learning to describe accurately what is seen; and (4) promptly recording what is observed. Although observations may begin without a particular question in mind—soon what is observed will raise questions, and ones that appear to have a chance of being answered can be pursued, scientifically. Children, in particular, are often very insightful in their questioning.

A Teenager's Experience.

In 1966, while I was engaged for several months in field studies under the auspices of the Australian National University in Canberra, Australia, our teenage daughter Mary, wanted something to occupy her time. I suggested she study the Australian Meat Ant (*Iridomyrmex purpureus*). Being a young naturalist type, she accepted the idea. She found a thriving colony almost in the shadow of the headquarters of the Commonwealth Scientific and Industrial Research Organization (CSIRO), the top governmental research agency in the country. Study of Meat Ants in the midst of the colony is a major challenge and risks many bites and stings. I'm surprised she bought the idea!

In the course of her study, which lasted about a month, she found some ants at the colony nesting site that, for no apparent reason, were pinned to the ground, struggling to free or right themselves, yet she could not determine at first what prevented them from breaking free. They died or disappeared without apparent cause. She persisted in her observations.

Finally, she decided to scoop up the soil beneath trapped ants to see if something was holding them in place. As a result, she found a slender larva that lay hidden in a vertical burrow, positioned with its jaws near the surface, from which it could ambush passing ants. She found many of these predators in "silo" death traps scattered among the tunnels of the ant colony. She had discovered a potent Meat Ant predator.

DEVELOPING AWARENESS & EXCITING INTEREST

The CSIRO staff was alerted. The Meat Ant is a species of considerable scientific interest to Australian biologists and agriculturalists and has been studied extensively. However, the local agency was unaware of the identity of the predator. They decided to rear the larvae to determine species identity.

Sometime after our return to the U.S., we received a beautiful colored photograph of the adult culprit—a Tiger Beetle (Cicindelidae).

This story emphasizes the importance and reward of careful observation. Our daughter knew little about Meat Ants at the start of her study. It was her persistence and attention to detail that did the trick.

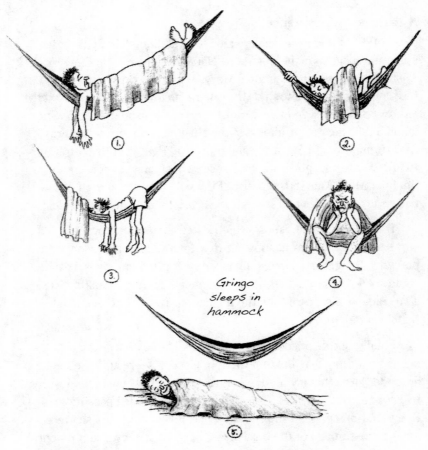

Figure 3. Sketches made on trip to the Sierra de la Macarena

Note Taking and Writing Skills

The Notebook

A notebook will help in conducting studies, because memory cannot be relied upon. Although there are many formats for note taking, I describe here some procedures established, over many decades, at MVZ. They have a proven track record. Pertinent are the Journal and Species Accounts (See Figs. 3 and 4).

The Journal

The journal includes where and when a trip afield occurred, participants, summary of observations and experiences, and perhaps photographs and/or sketches (See Fig. 3). The process can be like keeping a diary.

Species Accounts

Species accounts contain information on each species observed—a new page for each new species. It includes the species common name (or a tentative coined name, if the correct taxonomic name is unknown), followed by the scientific name, name of observer, locality of observation, and date and time (See Fig. 4). However, each following page should bear the name of the species, observer, and date if it is to continue from the preceding page. This is important should pages get separated.

To ease finding pertinent data, localities and plant and animal common names can be underlined with a wavy line, and scientific and other names of interest with straight lines.

A loose-leaf format makes possible keeping all notes for the Journal and Species Accounts (plant or animal, or other topic together).

At the end of a note-taking episode, pages can be assembled with all those pertaining to the Journal, followed by those for Species Accounts, in proper sequential order.

DEVELOPING AWARENESS & EXCITING INTEREST

Journal

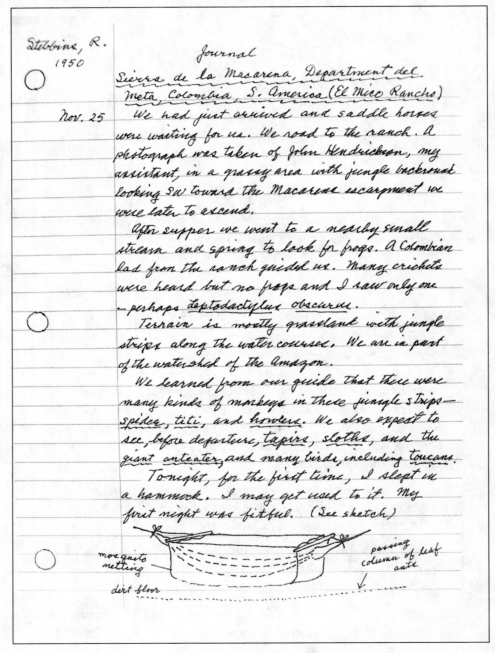

Figure 3. Format for Journal.

DEVELOPING AWARENESS & EXCITING INTEREST

Species Account

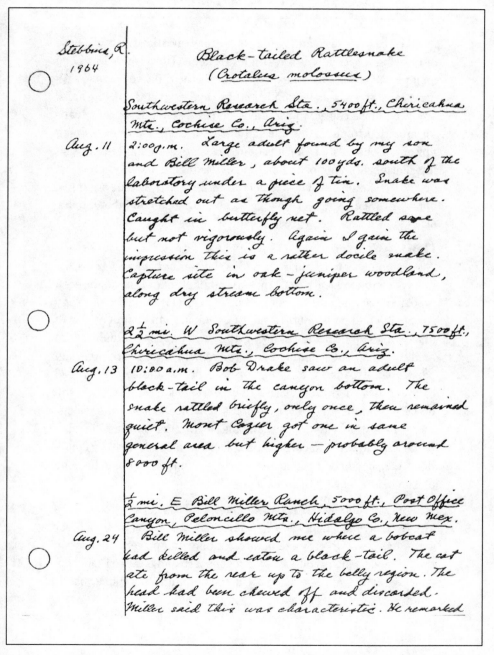

Figure 4. Format for Species Account.

A Case for Nature Drawing

The case for drawing: Some people may question the usefulness of drawing, especially if they find it difficult. However, it has been an integral part of human expression from the time we became self-conscious as a species, manifested in cave drawings, pictographs, and even maps carved in soft stone. Representational drawing calls for attention to detail and therein lies its great value in developing powers of observation, for it requires close examination of the subject. Formerly an important method of teaching in biology, it has lost ground. It should be revived to help restoration of the whole organism approach (See "<u>Nature Drawing</u>" Leslie, 1995), a book that emphasizes the value of the <u>process</u> of sketching, less so the outcome.

Getting Started

A good way to begin observational training is to write descriptions of common local plants, birds or other animals that can be seen repeatedly and at close range. Emphasize precision in terminology (See p. 36) and statement. To aid this, students can be provided with photocopies of drawings showing the terms applied to subjects of study, such as leaf structure and arrangement, external regions and parts of a bird, insect, etc. (Fig. 5) Students should be encouraged to make drawings that clarify their descriptions. Binoculars will be helpful in the study of birds and a hand lens for small subjects. In recording information on animals, focus first on identification, behavior, and time. Recording date, locality, habitat, weather conditions, etc., need not be done immediately, although it should be done on site before leaving the area if possible.

An introductory activity for young children. Most youngsters love to draw, even at an early age. I have used drawing in an outdoor setting as a way to promote nature-bonding and powers of observation. Provide each participant with a clipboard and 8 ½" x 11" sheet of paper, with the paper held in place at the bottom with a rubber band.

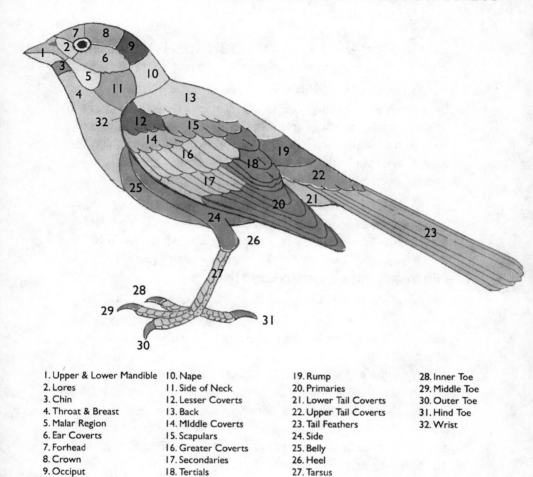

Figure 5. Parts of a bird.

In a sketching episode in Tilden Park, the subjects were ducks in a pond in the Nature Area. I urged the kids to look closely at their selected subject a while before starting to sketch. It would help them to make a more accurate drawing.

Sketching mobile ducks was a challenge. One youngster provided a male Buffle Head with an extra eye having failed to see the bird's dark eye, hidden in the patch of black on its head. She wanted to make sure her duck wasn't blind, so placed it in the white patch on the bird's head. The mistake drove home the primary purpose of the exercise.

DEVELOPING AWARENESS & EXCITING INTEREST

Science and the Search for Meaning

As observational skills develop, especially with older children, the search for meaning should follow, through application of the scientific method. After careful observation of a subject of interest, decide on a question, collect facts that seem to pertain to it, propose a hypothesis, and subject it to testing. Don't make the question <u>too general</u>. Nature can answer only "yes, no, or maybe," so be specific.

A scientific approach usually proceeds through trial and error—a step at a time. I sometimes urged my students to take considerable time to carefully ponder a question they wished answered before they gave it a title because the title would focus the goal, and if carefully considered would save precious time.

Often the hypothesis may fail and you may have to recast it or gather more facts. As the famous zoologist Thomas Huxley has said—"The great tragedy of science [is] the slaying of a beautiful hypothesis by an ugly fact." But keep trying. Always maintain a healthy attitude of skepticism. If the facts seem to be taking an unexpected direction, be guided by the facts. Such an experience is of value in the general education of everyone, for the scientific method and mode of thought is needed in dealing with numerous problems that confront us in all walks of life. Scientific thinking and methods do not belong only to scientists.

Developing Accuracy in Observation and Description

Careful observation and description is essential in all scientific work and is necessary for success in the affairs of daily life. Whatever the subject, accurate transmission of information requires commonly understood descriptive terms. Our subject will be leaves.

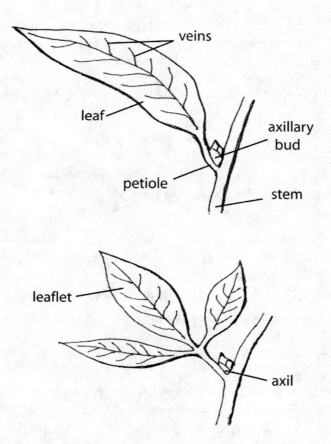

Fig. 6. Definition of a leaf here illustrated with a simple and a compound leaf. A leaf is an appendage of a plant with a bud at the axil. The axil is the recess between the petiole and stem. When the bud is not clearly seen, check dead leaves on the plant or on the ground below. The leaf-like appendages of a compound leaf are called leaflets.

Activity: Rediscovering the Leaf

The following exercise emphasizes these points as they pertain to work with living organisms. Leaves have been chosen as an example for study because of their availability and familiarity, and because many have features of external structure that are easily seen and can be sorted into well-defined categories. For the exercise to serve its chief purpose, participants should be unfamiliar with general leaf structure and variation. However, parts of the activity may be of value to more informed persons as well. It starts with students describing leaves in their own words, which is often vague, but later, after learning some helpful terms, they discover the value of more precise terminology. Many technical terms can be used (See Hickman, 1993). However, for our purposes a simplified version is offered. (Fig. 7, p.40)

This activity has been designed for the classroom and uses plant samples brought in for the lesson. The assorted samples of different species are placed on sheets of newspaper on the floor around the sides of the room. Each site can be imagined as a natural assemblage of species growing together in a small area outdoors.

The sites should be well separated and <u>numbered conspicuously in consecutive order</u>. Students will be visiting the sites to identify species based on student descriptions.

In choosing plant samples for this activity, do not use grasses, sedges, or coniferous species—things may get too complex. Check with local plant specialists to avoid any toxic species such as poison oak or poison ivy and milkweed or other species with white alkaloid sap. Nettles and sticky species should also be avoided. When collecting, carry a copy of Figure 7 with you so as to include many samples like those shown in the figure. Hopefully, a roving naturalist or laboratory assistant will be able to help with this program (See p.198—Jim Grant).

DEVELOPING AWARENESS & EXCITING INTEREST

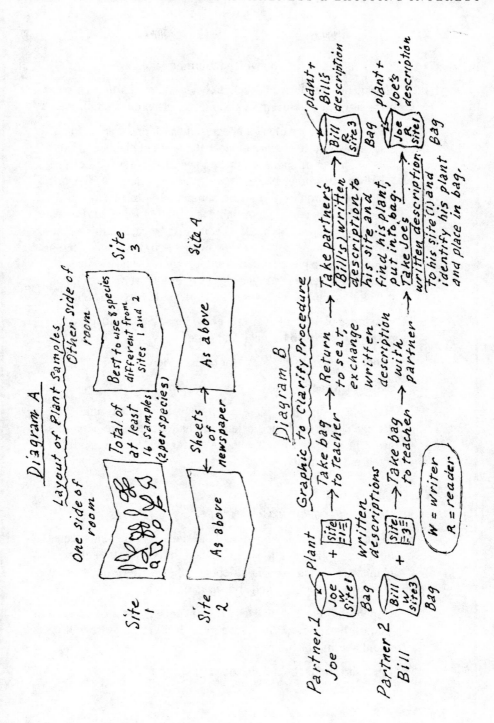

EQUIPMENT

1. Felt pens and papers on which to number sites

2. Newspaper (two attached sheets), at least four sets, to put plants on at sites (See Diagram A, p.37) under classroom set up)

3. At least two pieces of each species of plant, and a selection of 8 species or so (therefore at least 16 pieces of plants per site). The same set of species can be used at each site if desired but diversity among sites may add to student interest. To work well, plant samples should be somewhat similar. No highly unusual or obviously identifiable leaf should be used. Pieces of the same species should not be next to each other at the plant site. The pieces should be large enough so that when a student takes a piece from one sample there is still a stalk with a number of leaves left on it. Remove any flowers or fruits on the plant samples. The focus is strictly on leaves and stems.

4. Two large paper bags per student. The bag must obscure the plant so it needs to be large and paper.

5. Pencil, paper, and clipboard for each student

6. Scissors at each site to cut pieces of plants

7. Handout of terms (Fig. 7, p.40) to use in describing plant samples to be explained briefly by the teacher. The handouts are to be given to the students only after the "First Episode" of the present activity.

CLASSROOM SET-UP
(Diagram B helps explain the following process)

1. At each site, mix up and spread out the 16 or so plant species on the newspaper. The sites can each have the same array of plant species, as each is independent, but there must be at least two samples of each species at each site.

2. Depending on class size set up 2 or more plant sites on each side of the room not too close to each other as there will be about 3 or more students per site.

3. Each site should be numbered (1, 2, 3, etc.)

PROCEDURE *(Around an hour is required for this exercise)*

First Episode

1. Divide the class into pairs with one member of each pair being a number 1 and the other a number 2.

2. Taking a paper bag and a clipboard with a piece of paper and pencil, all the 1s go to the plant sites on one side of the classroom and the 2s to the other side.

3. All students select a plant sample, snip off a portion, including the stem with one or more leaves, write a careful description of their sample, and record the site number from which the plant came. They then put their sample in the paper bag and write on it their name, the site number, and a "W" for writer.

4. They now take their paper bag to the teacher, while retaining their description, and collect a second bag. They then return to their seats rejoining their partner.

5. The partners now swap their written descriptions and each goes to the site number written on the description to try to locate the plant described by their partner. Once they select the leaf sample they think is the right one they put it in their paper bag along with the written description. They write their PARTNERS name on the bag, and site number, and an "R" for reader and then take the bag with sample and description enclosed to the teacher.

6. The teacher keeps all the bags in order so that the bags with the same names on them are together.

7. Now comes the unveiling! The teacher (having the partner's bags next to each other) opens the bag of the writer (W) first and takes out the plant. He or she then opens the partner's bag and takes out the plant piece for a comparison in front of the class. (No names are required here so that no students are embarrassed. The names are simply for the teacher to make sure that the corresponding bags are opened.) Some written descriptions could then be read out to see how they could have been improved or what good descriptive words were used.

In a large class not all descriptions need be read right away. However, the teacher should quickly continue to open each pair of bags to see how many readers (Rs) were able to accurately recognize the plant sample by their partner's description. The total number of correct and incorrect determinations should be recorded for class discussion.

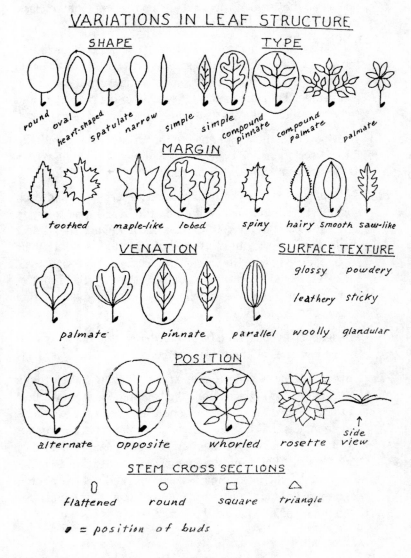

Fig. 7. Terms to use when describing leaves.

8. A brief discussion follows about how to accurately describe leaves including the ideas below:
   a. Relative meanings of words such as "big", "small", "long", "short" and how specific estimation of size would be more accurate (e.g. 2 inches long).
   b. What constitutes a leaf?
   c. Consideration of aspects of a leaf that should be described (e.g. Color, texture, hairy or not, type of edge—smooth, spiny, lobed, difference between upper and undersurfaces

Each participant should now receive, for the next episode, a handout explaining what constitutes a leaf (Fig. 6, p.35) and terms to use in describing leaves (Fig. 7). Students are to consult these resources as the teacher defines a leaf and explains some of the terms on the handout—such as ovate (egg-shaped), spatulate (like a spatula), pinnate (like the veins of a feather), and palmate (like the spread fingers of a hand). The position of leaves on the stem should also be noted.

Second Episode
1. After the discussion, and now with handout in hand (Fig. 7 on their clipboards), the students repeat the procedure, but at a different plant site and choosing a different species. This time, as the teacher reads out the descriptions, there should be a noticeable improvement in the accuracy of describing. Total up correct and incorrect determinations, and compare with the previous exercise. Keep a record of findings for future classroom comparisons.

To Take Home

Suggest to students that they make an album of drawings, photographs, scanned images or pressed leaves found about their homes or in the field with species names as they are determined. (See Stuart and Sawyer, 2001, for California). Among the leaves observed, find examples that illustrate the variation in leaf structure shown in Fig. 7.

Schoolyard Nature

Given increasing urbanization, an effort should be made to bring nature study directly to schools, especially where other alternatives such as cooperative study sites (See p.81) do not exist. Special focus should be at the <u>elementary and middle school levels</u> because (1) life-long attitudes develop early, (2) many children worldwide have no schooling beyond the elementary grades, and (3) in general, school curricula are more flexible than later. Activities that follow are mostly for children of upper elementary and middle school age, but teachers, parents, and caregivers may find some activities suitable for higher and perhaps even lower grade levels.

The primary goal is to develop an <u>ongoing</u> inventory of the plants and animals in the schoolyard and some of the ways they interact with each other and their environments. With animals, it should distinguish among native, transient, and feral species and, with plants, native, cultivated, and weedy species.

The objectives are to be achieved by a codified program of note-taking done by the students, with guidance from their teachers and perhaps at times with the help of interested parents or other caregivers permitted to be on school grounds during after school hours. Observations are to be recorded on an 8.5" x 11" formatted data sheet (Fig. 8) on the back of which is a stapled map of the schoolyard (Fig. 9, p.48). A descriptive approach will be used in documenting information on species encountered. In the course of a class involvement, some students may wish to be quite actively involved and time may be available to do so. However, I would hope, as a minimum, each student would complete at least three species account reports, plant or animal.

DEVELOPING AWARENESS & EXCITING INTEREST

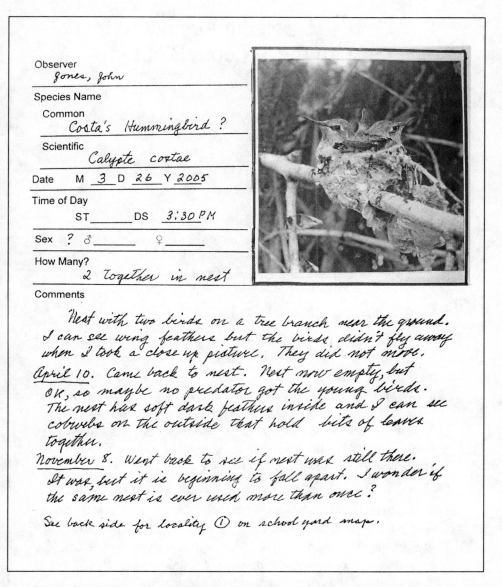

Observer: Jones, John
Species Name
 Common: Costa's Hummingbird?
 Scientific: Calypte costae
Date: M 3 D 26 Y 2005
Time of Day: ST____ DS 3:30 PM
Sex: ? ♂____ ♀____
How Many? 2 together in nest
Comments:
Nest with two birds on a tree branch near the ground. I can see wing feathers but the birds didn't fly away when I took a close up picture. They did not move.
April 10. Came back to nest. Nest now empty, but OK, so maybe no predator got the young birds. The nest has soft dark feathers inside and I can see cobwebs on the outside that hold bits of leaves together.
November 8. Went back to see if nest was still there. It was, but it is beginning to fall apart. I wonder if the same nest is ever used more than once?

See back side for locality ① on school yard map.

Fig. 8. Species data sheet: To be stapled or photocopied and attached to the back of the map of the school yard (See Fig. 9, p.48). It is to be used in recording information on flora and fauna observed on the school grounds. Note that the map shows the location of observation. Data sheet size to be 8.5" x 10".

Activity: Getting Acquainted with the School's Biota

EQUIPMENT
1. Map of school yard (8 1/2" x 11"), on the other side of which are directions for recording observations (Fig. 8)
2. zoom equipped digital camera
3. color scanner
4. camera printing dock for quick processing of photographs in color (optional)*
5. drawing materials
6. file cabinet for storing data sheets and/or a computer for digital storage and creation of a CD data file
7. species identification guides (housed in school library and checked out as needed.

MAP MAKING

Many cities have aerial photographs or maps of school grounds for use by police, fire, and public safety departments or perhaps it may also be possible to access satellite imagery or architectural drawings. Further, a basic map can be made on-site by teachers and students (See Hancock, 1991). However, if facilities and motivation are not available to do so, the program should <u>still go forward</u> without the detailed locality information <u>precise</u> mapping would provide. The school ground inventory can still proceed.

On the map of the schoolyard, including all areas not occupied by buildings—lawns, plant-beds, play fields, etc., create a grid to correspond to 8-yard intervals and number the sectors consecutively. Various readily-identified—"fixed" landmarks—trees, bushes, playground equipment, benches, drinking fountains, etc.—should be shown to aid in referencing locations of observations.

Compass direction should be shown and scale and date of issue should be indicated**. For a large schoolyard, enlarge the basic map

---

*Some photo shops will develop digital camera pictures at low cost.

** This is important because updating of maps will be required over time as vegetation and other physical changes occur.

and divide it into separate clipboard size (8 1/2" x 11")sections and numbered sequentially. Students can then be provided with photocopies of each section as needed for their studies.

Use of Maps

Maps make it possible to show accurately (1) the location of plant and animal species observed on the school grounds, (2) animal movements, and, if pursued over the long term (3) changes in the biota over time. In studies of animal movements, routes taken in foraging, nesting, sheltering, etc. can be recorded. The locations of bird and ant nests, spider webs (clearly seen when covered with dew or lightly sprayed with an atomizer), can be shown. For long term studies focus on selected, frequently seen, species. Such species may serve as indicators of environmental trends.

Data Gathering

Insects (ants, beetles, butterflies, caterpillars, etc.), spiders, snails, slugs, birds, frogs, salamanders, lizards, snakes, etc. can be photographed on the spot as can leaves, blossoms, and fruits (in botanical sense, including seeds, nuts etc.) and the pictures attached to the data sheets. Some can be scanned. In the absence of cameras, copies can be made of species shown in field guides, or students can make drawings.

Teachers and/or assigned student photographers may sometimes be able, on short notice, to take pictures outdoors if special opportunities arise—such as arrival of an unusual bird, beetle, or butterfly. Some subjects (such as invertebrates or plant samples) can be brought into the classroom, sketched, photographed, or scanned and the animals returned to marked capture sites. Even small dead animals can be recorded. Scanners have remarkable depth perception.

For schools without cameras and other equipment mentioned, students can make drawings of plant samples and some small captive animals such as tadpoles, frogs, lizards, etc. that can be housed temporarily in glass containers. Some schools may also have access to mounted specimens (birds, etc.) that can be suitable subjects.

CREATING A FOOD WEB *(See also Fig. 18, p.75)*

As information pertaining to the school's biota is gathered it should be possible to display a food web (who eats or interacts with whom), as it exists in the real world of the school grounds. Photos of the animal and plant members of the web can be enlarged and arranged to show the flow of food-derived energy from the sun via the green plant base, then through increasingly larger animals to the top of the food chain. A poster can be prepared for classroom or school display. Student teams can work on different species. If insects are brought in for study, mouthparts, and other structures can be examined with a hand lens or dissecting microscope to help decide their role in the ecosystem (See p.74).

Species Identification Files

Many guidebooks are now available for identifying species of plants and animals and universities, public museums, and specialists (hopefully a roving professional naturalist?—See p.198) can be turned to for help. For species that have not been identified, a temporary "coined" name can be used in the meantime. Students can imagine they are explorers who have found a new species, and must now describe and name it. Here is an opportunity for the teacher to introduce the subject of taxonomy and the importance of agreed upon scientific names for organisms.

Why is it so important to go to great lengths to ensure accurate information on nomenclature? It is because the name is the "handle" to what is known about the organism—not only from the standpoint of our intellectual interests in understanding its relationships and place in nature but also from practical concerns over its possible direct interactions with us. Example: It is highly important that we be able to recognize dangerous pathogens and venomous or toxic species and to clearly recognize known beneficial ones that may be confused with look and/or act alikes. This leads to the subject of mimicry—a harmless king snake with a color pattern that "copies" that of a dangerous, highly venomous, coral snake and gains protection thereby.

Setting Up the Files

Arrange the growing files of data sheets including photographs in proper taxonomic order. Species awaiting precise identification should be kept in a separate location until final identification is made.

Since students and teachers come and go it would be desirable to have the accumulating information on a school's biota, organized in long-term central storage available for call up for studies.

The schoolyard itself as a voucher file.

With established plantings of trees and shrubs the schoolyard itself can act somewhat as a voucher file. As identifications are made of these plantings, species identification markers, giving common and scientific names can be placed on or next to each plant. Indicate whether native or introduced and for the latter, its origin. The markers should be weatherproof and firmly placed in a manner done at botanical gardens. Students can adopt a plant by tagging it with their name and can keep notes on insects, birds, and other life that interact with their plant. They can check their plants during recess and on other occasions.

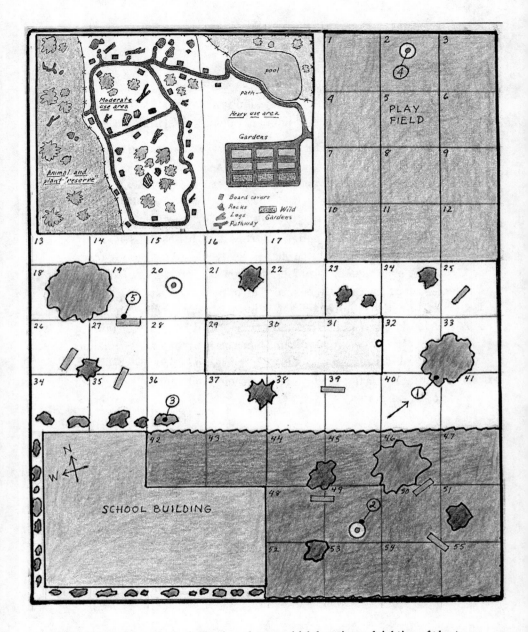

Fig. 9. Map of hypothetical school yard upon which locations of sighting of plant and animal species have been recorded (circled numbers). The numbered grid (8 foot squares) aids positioning of sightings, as do benches, drinking fountains and other fixed objects. A nature area has also been established. (See p.49). Actual map size would be 8.5" x 11".

Activity: Creating a Nature Area

In the 1950's Herbert Wong, principal of the Washington Elementary School in urban Berkeley, California, felt so strongly that children needed to connect with nature, he converted a large section (1/3 + ?) of the schoolyard into an area for nature study and gardening. During the decades of its existence, many Berkeley children were introduced to nature.

Features to consider in enhancing an area for school-ground nature study follows (See Fig. 10):

1. A fence with gate access that can be locked as needed.
2. Pathways to prevent damage to plantings.
3. If conditions permit, create a transient pond, dependent upon rainfall (See p. 226, Vernal Pools). In some areas, pools may attract mosquitoes and other pests. This must be considered in creating a pond. Check with authorities for methods of control such as introducing mosquito fish, which can be removed prior to pond drying, or native back-swimmers (notonectids) that will leave on their own. Take care in handling because they can deliver a painful bite.

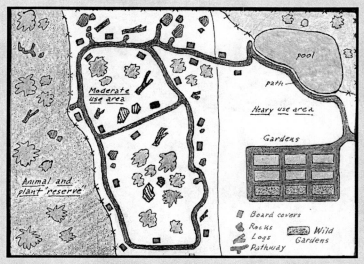

Fig. 10. Ground plan of a school nature area with both wild and cultivated garden plots.

4. Small garden plots, some unmanaged except for watering, for study of untended growth (See New Zealand cabbage patches program, p.74). However, now with food shortages increasing internationally, school nature areas would do well to emphasize local food production. (See p. 64-72)

5. Hinged board covers to attract small animals (See Fig. 11). Rocks and logs as sunning sites and elevated locations for spotting prey, attractive especially to sun-basking lizards.

6. Hopefully, in less-developed areas, save an area of natural, or little disturbed habitat, which can be a place for special wildlife studies.

7. Board covers can be like a magnet in providing a sheltered location for, lizards, salamanders, small frogs (if ground is moist), insects, spiders, and other small invertebrates. Since they are so effective in attracting small animals, details of construction and use follow.

Fig. 11. Hinged board cover for attracting small animals.

Board covers can be made from a piece of 5 ply wood about 12 x 16 inches or larger. They should be painted with waterproof paint—and can be matched to surroundings if desired. A 1-inch dowel or discarded broom handle should be nailed flush with one end and also painted to protect against decay. The dowel (or broom handle) should extend 3 inches or so on each side, where it is to be staked in place, yet free to rotate*. This is to ensure hinge action so the cover is always returned to the same position after examining the site for animals. A drawer pull can be attached to ease handling. Covers should be numbered to keep track of animal visitors that are recorded in student notebooks. In a place of frequent use a rotating system of lock downs should be established so that each cover has an interval (several weeks?) free of disturbance.

Note the gradation in-space when the cover is in place. This allows animals of different sizes to wedge in—appealing to their thigmotactic (body contact) tendencies. When installing a board cover, scrape away grass and other objects to provide a relatively smooth earth surface beneath. You may be surprised at how effective such covers will be in attracting small animals that can then be examined and studied and returned to their cover site.

If covers are in areas where vandals may be a problem, attach a waterproof tag noting their use for scientific and school purposes.

See the National Wildlife Federation Backyard Wildlife Habitat Program that can also be applied to school properties (p.77).

---

*Hooked or u-shaped metal stakes suitable for this purpose are sometimes carried by hardware stores.

Activity: Live Animals in the Classroom

Developing nature-oriented goals can be greatly aided by having some live animals in the classroom, especially at the elementary school level. Just the sight of a live animal can often deeply stir the heart, especially of a young person. My U.C. Berkeley colleague, Professor Ned Johnson, said he became "hooked" on birds at the age of seven while in a park with his mother when a Red-shafted Flicker caught his eye when it lit in a tree four feet away. "It happened in an instant. I just couldn't believe how beautiful it was. That flicker just crystallized things for me." He became a world-renown ornithologist. There follows some suggestions, devices, and activities with which I have had personal experience that can support the classroom approach.

EQUIPMENT
1. Bird feeders
2. Berlese funnel
3. Small Aquarium
4. Aerator
5. Dissecting and compound microscopes (at least one of each) to be rotated for classroom use

THINGS TO DO
1. Place a bird feeder close to the window of your classroom. Include one for hummingbirds.

2. Captive-reared frogs, reptiles, insects, etc. can sometimes be obtained at vivaria or pet shops, thus reducing pressure on wild populations. Students may enjoy arranging for housing and care. Large numbers are not required. One or two species should suffice. Seek guidance from universities, nature guidebooks, zoos, or, hopefully, a roving naturalist (See p.198)

3. For purposes of identification and close examination bring in an insect, other invertabrate obtained in the schoolyard for inspection. At the same time there will be opportunity to watch behavior. If kept only briefly and obtained locally, it can later be returned to its marked capture site.

Many small invertebrates, including some flying species, can be temporarily slowed down for observation by putting water in their container and shaking them gently for about 30 seconds. They can then be allowed to crawl out on a pencil where they can be examined at close range with a hand lens. Grooming and other behavior can be observed. Re-immerse them as necessary, but don't over do it. Some may get away, so this activity is best done just outside the classroom. Have the students write in their notebooks what they observed. Small animals can be caught in clear plastic vials or jars. By holding the container in one hand and the lid in the other, a stationary or hovering insect, even a butterfly, can be caught by getting the animal between lid and container, then quickly popping the lid in place.

The Berlese Funnel

It is used to catch <u>small animals found in both leaf litter and soil and on the interface between—where the litter contacts the soil surface.</u> The interface is a crucial area in soil formation and the animals there are an important agent in returning nutrients to the soil.

To make a Berlese funnel obtain a desk lamp, a jar, and a large, wide-stem <u>opaque</u> funnel (like that used to drain crankcase oil). A hardware-cloth shelf (1/4 inch mesh) is cut to fit and placed across the middle of the funnel. Two arrangements are shown. Figure A uses a collecting container with 70% alcohol for quick preservation of specimens for detailed, more advanced studies. Figure B uses as the receptacle, an opaque vial around 2 inches in length or longer by 1 inch diameter, containing a tuft of moist cotton to prevent desiccation and held in place with a cotton seal around the stem of the funnel. The vial should rest in the center of the empty supporting container.

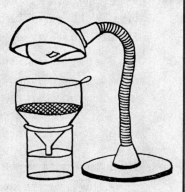

Fig. 12A. Berlese Funnel with 70% alcohol for quick preservation of specimens.

Fig. 12B. Berlese Funnel for studies with live subjects that can be released.

Place either a soil or leaf litter sample in the space above the shelf. Turn on the lamp (25-40 W) and lower it over the funnel container, but don't get too close. Just enough light and heat is required to drive the small animals into the container below. If the heat source is too hot and close to dried leaf-litter, the litter might ignite. Thus with dry leaf litter caution is especially called for. A soil sample may take several hours for the animals to collect in the jar. Leaf litter, less compact, will take less time. With the live animal approach, place a layer of moist cotton in the jar to prevent desiccation.

Source of Leaf Litter

Find an area of dead leaf-fall. Scrape through litter until you find a site that shows a good level of small animal activity. Sites where the litter has been undisturbed on a damp soil surface are usually best. Protect your hands with gloves. Avoid areas sprayed with pesticides.

Put the litter in a wide-mouth gallon jar and close with lid, perforated to allow air flow. Since the space for litter in the Berlese funnel is shallow, to get a good take of small-animal life may require several refills of the funnel receptacle as each refill dries out. Make prompt use of samples collected.

Bush-shaking and use of a tray

A tray is pushed into position beneath an area of leafy branches and the branches above shaken or struck to dislodge small animals that then fall onto the tray. A double thickness of newspaper sheet works well. The relatively smooth surface of the paper and flexibility allows it to be bent into a trough and to shake its contents into a wide-mouth gallon jar. If you have an insect net, it can be used. (See p.148) Draw the bag to one side against the net rim to make a tray. With the other hand tap the bushes with a short stick to knock small prey onto the net.

Use of Animal Samples

Samples of captives can be transferred to capped Petri dishes, where they can be examined closely under a dissecting microscope or hand lens. Placing the Petri dish on crushed ice will reduce somewhat the mobility of captives.

For younger students, below fifth grade or so, simply seeing the diversity of creatures will be an important lesson in itself. Hopefully, they will especially come to understand that soil is more than "just dirt." It is a complex living system.

For more advanced students, urge species identification and study of anatomy, including mouthparts and other anatomical features, to try to determine the role of species in the ecosystem. As noted, sample specimens can be preserved in 70% alcohol. The effect of such sampling would be insignificant in comparison with the great damage we are inflicting on such organisms in so many other ways, and the learning experience can be profound and contribute to a respect for nature. Immobile individuals are needed for drawing. Live animals not used in study can be returned to marked capture sites.

The Importance of Soil and the World of Small Living Things

Awareness and understanding of the small world of living things, often overlooked, should be high priority in the education of everyone because of its role in maintaining the fertility of soil and in natural recycling—processes upon which all life, including ours, depends. Unfortunately lack of understanding, or indifference in this area, can result in high environmental costs.

Most virgin soil swarms with life. A corporate farmer in the Great Valley of California once lamented to me that the soil he depended on for his crops was essentially a dead zone. Decades of use of pesticides and chemical fertilizers had essentially destroyed the soil community. He worried about the effect on the nutritional quality of his crops.

In natural recycling soil organisms are crucial to converting constituents following death for reuse by new life. This is a world of bacteria, fungi, lichens, protozoa, diminutive insects, and other small invertebrates.

I remember the excitement I felt as a child when my Dad placed a drop of water on a compound microscope slide and asked me to have a look. The tiny drop swarmed with amazing translucent creatures of many shapes and sizes, vibrating with life. The source was dead grass that had been immersed for a week or so in a jar of water. Even though the names of the creatures were unknown to me, their images have never been forgotten. Teachers should arrange for such a display.

DEVELOPING AWARENESS & EXCITING INTEREST

Frogs and their Life Stages

Frogs are so called "cold-blooded" animals. However, they are better called "ectotherms" because their body temperature is derived from their surroundings and is controlled chiefly by movements to and fro within the thermal mosaic of the environment. Ectotherms include all animals other than the birds and mammals. The latter are referred to as "warm-blooded", but are better called "endotherms" because their body temperature is generated from within and is under metabolic control. Some shifts in control of temperature, however, occur with fever, and in some species with hibernation and other metabolic demands.

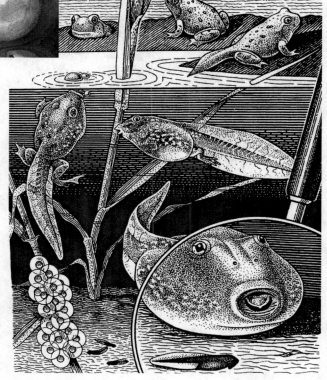

Fig. 14. Developing embryos of a clouded salamander (Aneides ferreus) seen through the transparent glasslike egg capsules common among the amphibians. Photo by William Leonard.

Fig. 13. Life stages of an "amphibious" amphibian, the Western Spadefoot Toad (Scaphiopus hammondii). Note eggs with transparent capsules.

Activity: Studying Frogs and Tadpoles

Most frogs are especially good subjects for connecting with nature and it is possible to study them without undo damage to wild populations.* Just a few eggs and/or young tadpoles can be brought in from a local pond to start things going, and you can be sure students will be captivated by what happens. No elaborate equipment is needed for care. However, each classroom should have access to a dissecting microscope for close-up viewing of changes in life stages. (See Fig. 13).

Many amphibian eggs, including those of frogs, have transparent gelatinous coats, allowing direct viewing of embryonic stages—in a continuum to the time of hatching. (Fig. 14) For a time, until their mouthparts fully develop, the hatching larvae (in this case tadpoles) are nurtured by their yolk. When ready to forage many species can be fed algae from their pond site. Alternatively, I have found a diet of lightly boiled lettuce, and crumbled bits of hard boiled egg yolk for protein needs, works well if attention is given to keeping the water fresh by frequent changes and not allowing feces and rotting food fragments to accumulate. Ideally, water from the pond site should be used. As the time for transformation to the froglet stage approaches, provide floats or above water rock surfaces for the froglets to climb out on. They will require live small animal food. Small animals obtained by use of the Berlese funnel could feed them or pet stores might be a source for food.

---

*Some salamanders are equally good subjects but their larvae, soon after hatching, will require small live animals for food. However, algae and other plant sources can be obtained at pond sites that often contained associated small animals that can be shaken into the water containing the developing larvae. Some pet shops may also be a source of prey. Check with local wildlife agencies to determine if there are any restrictions on taking local amphibians.

It is important to avoid crowding in arranging an aquatic site for care. For small species I usually allow 2 to 4 tadpoles per half gallon of water and use a container that can provide a large water surface in relation to depth (also an aerator can help). A glass container, approximately 12 inches wide with sides 4 or so inches high has works well. Alternately, an aquarium can be used, and tadpoles at various life stages, can be transferred with a tea strainer to a water-filled dish for examination.

At the elementary grade level, during the embryonic development phase of the froglets, it may be enough to have the children note obvious changes in embryonic growth—when the eyes become pigmented and conspicuous, heart beat and blood flow begin, limb buds sprout, and fingers and toes start to form. Children can be greatly excited as they see the heart start pumping. It may be the beginning of respect for life—including their own?

Heart Rate of the Embryo Froglet

What is the heart rate of the embryo froglet seen through its transparent gelatinous egg membrane? Does it change with temperature of the water (check with thermometer). Add warm or cold water and avoid prolonged temperatures above around 18-20°C. Allow time, in each case, for temperature stabilization. Then check heart rate and water temperature. Graph results.

Marking Wildlife for Field Recognition

The following stories demonstrate how relatively simple projects of marking individual animals for field recognition can create great interest. They illustrate the commitment young people can feel in a discovery mode of study. I believe it appeals strongly to humanity's ingrained love of the hunt.

Being able to recognize the individual members of an animal population makes possible obtaining valuable information on individual movements, interactions with each other and other species, growth rates and longevity, size of their home ranges and territories, mating behavior, etc. that otherwise would be unavailable. In some studies it also carries with it the feeling of revisiting old "friends."

A Spider Story

In the fall of 1967, a meadow in the Nature Area of Tilden Regional Park in the Berkeley Hills, California, contained a large number of orb-weaver spiders—the beautiful Golden or Orange Garden Spider (Argiope aurantia)[4] (P. 1). (Fig. 15) Among children bussed to the area for nature study were two boys in their early teens who had become interested in the spiders. They were impressed by their size, large symmetrical webs, and especially how the spiders caught and fed on their prey. I suggested that they mark the spiders for individual recognition and study what went on in the spider colony.

Using red paint* and a slender toothpick, the boys marked each spider with a small red spot on a distinctive location. A week into the study, spider no. 12 was found in a new web 8ft. from her marking site and a larger individual was either in her former web or had built a web of her own at the site.

I now had four boys on the project. (Interest was obviously growing) On their own, they had established a division of labor—one numbered and drove stakes marking web locations, another marked

---

*Acrylic paint is effective in marking. It dries rapidly and is long lasting.

spiders, another recorded notes, and the fourth plotted the location of webs on graph paper.

The boys now asked me to come at 8:00 AM. Ten to 12:00 is not enough time, they said. They were recognizing problems: how to move about the area without damaging webs and how to mark webs. The boys rejected my proposal that we take a break and go on a hike. They were too preoccupied with their project!

A rain occurred. Spiders were rebuilding webs. The boys were fascinated with the process and were recording notes. Two spiders were found building webs in clumps of sedge at the base of which were dense clusters of "hibernating" (and breeding!) ladybird beetles. Were they counting on a food bonanza? One web contained a ladybird wrapped in webbing.

Fig. 15. The Golden (or Orange) Garden Spider (Argiope aurantica). Photo by J. Patrick Li.

The public showed great respect for this project. The spider area was delimited by a fragile lath-stake-chalk-line fence upon which signs were hung, "Junior Ranger Spider Project. Do Not Disturb." The meadow was a favorite place for young and old park visitors, yet no one did any damage whatsoever to the project. Hundreds of individuals surely visited the meadow during its existence.

Spiders are ideal subjects for field studies because of their abundance, but the bites of some are dangerous so species identity must be known. Two common web types are the orb and the tray (See Note 4, P. 1 for more information).

An Ant Study Involving Group Comparisons.

Another approach to marking animals is illustrated by a field study on the Australian Meat Ant (*Iridomyrmex purpureus*). It was done as a secondary school research project by my granddaughter, Miriam Broadhurst, in Adelaide, Australia. In this case distinctive marking for every individual was not required. Worker Meat Ants are around 3/4 inch in length, so they are large enough to be marked, if detailed individual recognition is not required. In some areas they are considered a pest, thus, knowledge of their biology is important to agricultural interests. Factors involved in affecting population densities and spacing of colonies are of interest.

In this study two hypotheses were proposed: (1) The <u>residents</u> of one nest will react aggressively toward members of another and (2) the farther the nests are apart the more aggressive will be the attacks on <u>foreigners</u>, because they likely would be more divergent in characteristics (odor, behavior, etc.) and hence less recognizable. Three nests were studied in an approximately linear row, 10.5 and 17.5 m apart.

Ants were caught by allowing them to crawl up on a stick and were then knocked off into a container. They were cooled in a freezer for 3 minutes to immobilize them. A slender tipped paintbrush and white acrylic paint were used for marking.

Twelve ants from each nest A-C were marked dorsally—6 with a white spot on the abdomen (controls) and 6 on the thorax. Six "experimentals" (from an adjacent foreign nest) and 6 controls (residents) were tested, one at a time, at each nest in sequence.

Antagonism toward foreigners by residents was noted at all nests. There were three levels of increasing aggression by residents. The first consisted of raised abdomens and searching with antennae; the second—searching with antennae and dragging the foreigner from the nest by the leg, thorax, antennae, or jaws; the third—the foreign ant was <u>dismembered</u> and dragged out of the nest.

DEVELOPING AWARENESS & EXCITING INTEREST

Results on possible changes in aggressive behavior with distance between nests were ambiguous. Perhaps more nests would be required, spread over greater distances.

In working with ants at the site of their nests, boots should be worn and long pants should be folded tightly over the boots and held firmly by rubber bands. Some species pack a potent bite and sting.

There are often some significant limitations on methods of marking to consider. The process itself may adversely affect survival of the marked individuals if it spoils camouflage, interferes with species recognition, increases susceptibility to infection, impairs locomotion or other aspects of behavior. There is also the problem of loss of diagnostic marks with shedding.

Fortunately, the members of some species populations differ enough in coloration and markings to be individually recognized. Domesticated cats that might frequent a schoolyard, which may be of interest due to their role as predators, are and an example. Also, some patterned insects, amphibians (some spotted or blotched salamanders and frogs) and other native animals are sufficiently different in markings (at least as they mature) to make possible something like a fingerprint file of photos or sketches making individual recognition possible. Spotting in many salamanders can be related to the numerical position of the costal grooves along the sides of the body. Make a drawing or photo of the diagnostic spot. Consult local natural history museums, wildlife agencies, and universities for advice.

Even if markings are not long lasting, information of interest can be obtained before they disappear. With lizards, I have used water-soluble indelible pencil colors applied by moistening distinctive marking sites with saliva and rubbing in the color. Care should be taken to use colors that are visible at close range for identification, yet not so conspicuous as to disrupt camouflage.

DEVELOPING AWARENESS & EXCITING INTEREST

Gardens for Learning and Food

School gardens, used to teach principles of agriculture, were common in the United States early in the last century, but now have largely disappeared from the <u>predominant</u> public school curriculum. School gardens are common in some less industrialized countries, as in south and Southeast Asia, where I visited some of them. <u>They are needed now more than ever</u>—to bridge the widening gap between people living in cities and the living and physical sources that sustain civilization. Fortunately they are beginning to come back, and there is now hope that they can become an integral part of public education (See p.71, Growing Science Lessons). They are one of the best routes to nature-bonding and recognition of our absolute dependence on the natural world.

Since we may soon require a national and worldwide mobilization to cope with the growing disparity between the world's food supply and escalating human numbers, a national and worldwide gardening program involving the public schools, should be planned for and given governmental support (See World Watch 1996, Vol. 9, No. 6, "Urban Agriculture").

We in the United States can lead. We should revitalize something like the "U.S. Garden Army" (See below) and greatly expand urban and suburban agriculture and support of the family farm as was done during World War I, followed by a similar effort in World War II.

With our rapidly growing population, failure to adequately protect our farmland, and considerable reliance on uncertain food imports, we place our food supply at risk unless national attention is focused on this important matter. <u>Our government should make every effort now, to place all fertile farmland off limits to development</u> and encourage the reestablishment of a national school and home program to help our food supply.

DEVELOPING AWARENESS & EXCITING INTEREST

A shift in energy sources from declining fossil fuels to biomass (plant residues),* hydrogen, wind, solar and other energy sources, will be increasingly required to run the agricultural economy with its demands for fertilizer, water pumping, distribution, etc. We are, so to speak, currently "eating fossil fuels" because existing modern agriculture is so greatly dependent on this energy source. It has been estimated that some 400 gallons of oil equivalents are expended annually to feed each American (Pfeiffer, 2003).

In our country, "Urban sprawl and poorly planned development continue to destroy over a million acres of farmland every year—currently an average of more than 3,000 acres lost every 24 hours! And our best land—the land that yields close to 80 percent of America's fresh fruits, vegetables and dairy products—lies directly in the path of sprawl. The vast majority of farmers love the land that is the center of their lives. They're eager to do whatever they can to keep that land healthy (Grossi, 2003**)." To rely heavily on food from abroad, that may decline in availability and not meet our health standards, in an increasingly crowded world, is a prescription for disaster! Attention to our country's human carrying capacity (See p.179) is also crucial to this issue.

---

*However, unless our demands are greatly reduced there will be growing competition between agriculture and fuel needs—biomass desired for return to the soil to maintain fertility and that required for fuel.

** The American Farmland Trust, 1200 18th Street, N.W., Suite 800, Washington, D.C. 20036 is an important source of information on farming in the U.S., with special concern for support of the "family farm" and its relationship to wildlife and environmental protection.

Fig. 16. (Above) Children voting. (Below) Children banking their money at the "Garden City" headquarters on the UC Berkeley campus. The frame building was constructed through a community effort involving university students and the older children.

Learning from the Past—An Historical Background and Beyond

(See also Lawson, 2005, A Back to the Soil Movement?)

As a result of concerns over the movement of people from farms to the cities, a major "back to the soil" movement developed in the United States in the early 1900s (especially, it seems, during the period 1909-1924). <u>Principles of agriculture, nature study, and accompanying civic values</u> were to be taught to children in the public schools as an important part of public education.

During that time an agricultural program began at the University of California, Berkeley, which eventually spread throughout California and the United States. Professor E.B. Babcock, a plant breeder, of the University's Agricultural Education Division, who developed the Babcock Peach, and my father, Cyril A. Stebbins, head of the Department of Junior Gardening, were prime movers (See Stebbins, 1913; 1924).

During the above developmental period, under the direction and supervision of my father and students in his nature study class, an area on the Berkeley campus was set-aside for Junior Gardening. It soon was covered with over one hundred 6 x 9 ft. garden plots, separated by pathways, where children from the Berkeley schools (ages 6 to 16 years old) grew vegetables and flowers they sold to Berkeley residents. Each youngster "owned" a garden plot and the older children helped the younger ones. The children elected a mayor, manager of garden tools, two policemen to ensure peace and harmony in the small community, and other community leaders. A small building served as the communities' headquarters where the children banked their money from sales of their produce.

The goal was not only to teach principles of agriculture and nature study, but also the make up and responsibilities of community life and the process of self-government.

DEVELOPING AWARENESS & EXCITING INTEREST

"The Garden City" had a constitution patterned after the Berkeley Charter. The program was so successful it spread rapidly throughout the state and country. (See Fig. 17, letter to my father from Luther Burbank, famous American plant breeder and horticulturist.)

Then came World War I, starting for the United States in 1917. The food issue for the country loomed large. "Food will win the war" was the battle cry. Fortunately, by now gardening programs had spread widely throughout the country and were well established.

A national School Garden Army was formed and my father, an intellectual leader of the gardening program, was at the helm for the western states (California, Oregon, and Washington) with headquarters at the Ferry Building in San Francisco. To support the war-gardening program, seed companies donated seeds for the cause.

I remember well, my father's stationery, some of it preserved in a family file—A crossed rake and hoe at the top, and beneath, "U.S. Garden Army."

The "Victory Gardens" of World War II benefited from the World War I experience and the agricultural training that became a required subject in many elementary and high schools in the 1920's. With the passing of the wartime emergency, the growth of corporate agriculture, and the decline of the family farm, school and home programs languished. Let us now ensure that they are fully revitalized. Many positive programs are underway (See Halweil, 2002).

The Values of Local Food Production.

Private and communal gardening programs, rooftop gardens, and conversions of lawns and other plantings to grow food should be encouraged (See Brown, 1995; and Halweil, 2005). North America's lawns occupy more land—about 32 million acres—than any other "crop" including wheat, and the average American is said to use ten times the chemical fertilizer and pesticides on a lawn than the average farmer does on crops (See Population Connection—The Reporter, Vol. 35, No. 3, 2003). People today, where possible, should prepare to grow some of their own food and our young people would benefit in learning to do so through school and home gardens. Such a program can not only help us deal with the prospect of growing food shortages,

DEVELOPING AWARENESS & EXCITING INTEREST

December 15, 1911

Santa Rosa, California
U.S.A.

Mr. C.A. Stebbins
Berkeley, California

Dear Sir:

I am delighted with your letter of December 14th and especially to know how well you are carrying out your project. I think it is the most important one inaugurated in California, or perhaps in the world.

I cannot imagine anything more useful or more interesting than your work. It is on the right line and you shall have my most hearty cooperation at all times when possible and my most hearty wishes for its success at all times. More than ever do I appreciate its great significance and value to the young people.

Faithfully yours,

(Signed) Luther Burbank

Fig. 17. Letter to my father from Luther Burbank

worldwide, but can, at the same time, be designed to increase ecological literacy. Communal gardens provide camaraderie and relieve the stresses of modern life. Furthermore, small community efforts are best able to use an ecological (organic) approach to agriculture. With the rapid increase now (2008) in the cost of food, it is gratifying to see the great increase in these communal efforts.

After field studies spanning 23 years, researchers at Cornell University, the Department of Agriculture, and the Rodale Institute have confirmed that the soils used in organic farming reduce emissions of carbon dioxide—the most common greenhouse gas. In conventional farming, strong chemical fertilizers rapidly break down organic matter, sending carbon dioxide into the atmosphere rather than keeping it in the soil. Furthermore, abundant activity by mycorrhizal fungi in organic soils also slows decay. Thus, less CO_2 is released, and harmful pesticides and other chemicals are avoided." (See Sierra, March/April 2004).

"Everyone depends increasingly on long-distance food. For some, this offers unparalleled choice, but it often stifles local cuisines and agriculture, while consuming staggering amounts of fuel, generating greenhouse gases, and compromising food security.

Fortunately, the long-distance food habit is beginning to weaken under the influence of a young, local foods movement. From peanut butter makers in Zimbabwe to pork producers in Germany and rooftop gardeners in Vancouver, farmers, start-up food businesses, restaurants, supermarkets, and concerned consumers are creating a revolution that can help restore rural areas, enrich poor nations, and return fresh delicious, and wholesome food to cities (See Halweil, 2002, Home grown, The case for local food in a global market—Worldwatch 163. See also 6 (P.1) 237).

DEVELOPING AWARENESS & EXCITING INTEREST

Gardens for Nourishing the Mind as Well as the Body

Excerpts from "Growing Science Lessons" by Robert Sanders

(Fall 2001): As gardens sprout in schoolyards around the country, a group of educators at U.C. Berkeley is intent on making sure they nourish the mind as well as the body.

"We're trying to make the school garden an integral and exciting part of the school's classroom curriculum," said Jennifer Meux White, associate director for education of the U.C. Botanical Garden and a specialist in science education. "The garden can be a rich learning experience."

In 1995, California's state school superintendent Delaine Eastin mandated "a garden in every school" to "create opportunities for our children to discover fresh food, make healthier food choices and become better nourished." Yet, little money was devoted to helping teachers make the best use of gardens, White said. "After the initial burst of enthusiasm, gardens would often languish and die because they had not become an integral part of the curriculum."

However, four years later, in 1997, two curricula had been developed and tested by more than 4,000 students, and today many teachers are using them across the state. One, "Botany on Your Plate," published by the National Gardening Association (NGA) in 2006, brought biological and cultural concepts into a series of activities and discussions of what we eat. The second, "California Habitats Alive!" focused on plant diversity. A third, published by NGA in 2008, "Math in the Garden," gave children and adults opportunities to practice and develop mathematical skills while delighting in the outdoors.

"This is a movement all across the country," White said. *"More and more evidence shows that getting children out of the classroom is a powerful learning strategy that takes advantage of their enthusiasm and interest. Out of doors, a lot of new observations and questions come up that reinforce and add to the lessons."*

Rebecca Burke, a sixth-grade teacher at Berkeley's Martin Luther King Jr. Middle School, has been using gardens in her classes for more than ten years—developing her own teaching materials and creating her own projects. Having worked with White for two years, she is enthusiastic about the ideas and concepts of the new curricula.

"Gardening is pretty infectious—it's easy to reach kids through such activities," Burke said. *"What's really great about Jenny's curriculum is that it helps me teach difficult concepts in ecology. Its hard for kids to understand how it's all connected—concepts like biodiversity and different habitats."*

King Middle School is known for its "edible schoolyard," supported by famed restaurateur Alice Waters (See McManus, 2004). But the curriculum works for schools without such backing, such as Lazear Elementary School in Oakland, where White has worked with second-grade teachers to create a raised bed and wine barrel garden over an asphalt area.

"We try to help each school have a garden it can support."

At Lazear, the cultural aspects of food and plants are particularly appealing to the primarily Spanish-speaking children. White was able to obtain from the U.C. Botanical Garden two seedlings of the sacred tree of Mexico—the hand-flower tree, flor de la manita, revered by the Aztecs and used in medicines for relieving pain and inflammation—for the kids to plant and care for. She also is providing mulberry trees to feed silk worms used in a popular classroom-teaching unit.

"We encourage schools to plant not just any plant, but to ask the faculty, community and families what they would like to grow," White said. "These gardens won't survive without teacher, community and parent support."

Gardens on school, home, and church grounds, and in both rural and urban areas, should now become an international goal.

A Chinese Proverb— Secret of Happiness

"If you want to be happy for a few hours, drink wine until your head spins gaily. If you want to be happy for a weekend, get married and hide away. If you want to be happy for a week, kill a tender pig and eat it. If you want to be happy all your life long…become a gardener."

Untended "Gardens" and the Teaching of Ecological Principles

For the nature-centered goals here expressed, it can also be useful to have some untended areas allowed to develop on their own, or deliberate plantings but left untended except for watering to maintain growth. Their purpose is to provide a place in the schoolyard (or if space is available, in one's own yard), where the development of an interacting, <u>unmanaged</u>, living system of plants and animals can be observed. In this way children can learn about unattended nature first hand and, if given proper guidance, come to better understand how natural ecosystems work, regardless of their size.

The New Zealand Primary School Program

(See Wellington, Department of Education, 1967)

This approach has been an integral part of primary school education in New Zealand. (See Fig. 18) A common practice has been to grow plots of cabbages, which form the plant base of a food web that soon develops. Students observed and recorded what animals came to the cabbages, which ones resided there, how they fed and sheltered, and how they interacted with each other and the cabbages. The garden soon had its herbivores (aphids, caterpillars), predators (insect-eating birds and spiders), parasites (parasitic wasps), and scavengers (slaters or sowbugs feeding on decaying leaf litter). Field trips to larger natural ecosystems occurred reinforcing and expanding on the knowledge gained from study of the garden plots.

A Few Explanatory Terms:

A food chain contains a sequence of species each of which is dependent on the preceding one for food. Food chains rarely contain more than four components. The last species in the chain is often known as the **terminal predator.**

Careful observation should enable pupils to construct such chains. Most species appear in more than one food chain. Such interacting food chains constitute a food web. (See Fig. 18)

DEVELOPING AWARENESS & EXCITING INTEREST

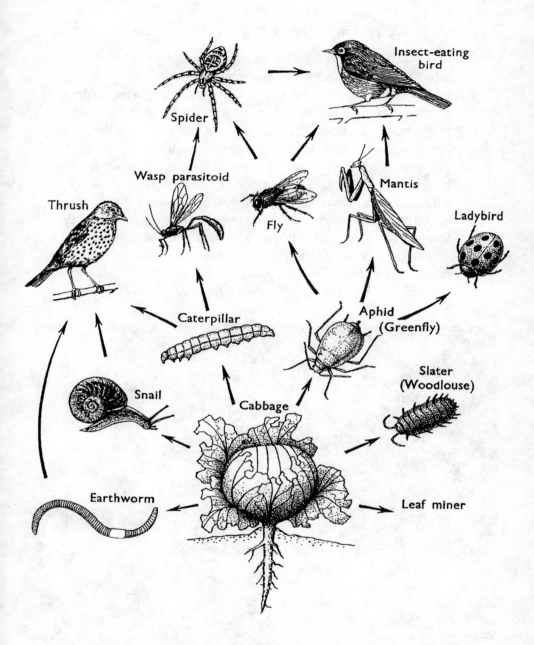

Fig. 18. Food relationships based on the cabbage used in ecological studies in the primary schools of New Zealand. (From M. Van Hove).

DEVELOPING AWARENESS & EXCITING INTEREST

Creating Home Nature Programs

A home nature area and food-producing sector have been established by Mr. and Mrs. Steve Abbors and their children in a suburban neighborhood in Walnut Creek, California. The tended garden area is fertilized with compost from house and yard refuse; help with pest control is provided by purchased lady bird beetles and establishing plants that attract native insect predators; pollination is aided by plantings designed to attract pollinating insects and by providing housing for native hole-nesting bees! (Fig. 19) Recycling of yard leaf fall is sped up by scattering bird seed among the dead leaves, which attract migrating sparrows in large numbers. The birds break up the leaves to find the seeds and fertilize the area with their droppings. The nature area contains local native plants, has an established colony of Western Fence Lizards *(Sceloporus occidentalis)*, occasionally Southern Alligator Lizards *(Elgaria multicarinata)* and is visited by

Fig. 19. (Above) Home garden and wildlife areas at the Abbors' residence in Walnut Creek, California. (Right) Apartments provided for hole nesting native bees.

many birds. The reptiles and birds help with insect control and the alligator lizards are known to include the highly venomous Black-widow Spider and its egg cases in their diet.

The vegetable garden is enclosed with plastic piping covered with see-through nylon netting with a mesh that allows resident fence lizards access for insect control. The weighted net side panels are attached to boards that rest on the ground, but can be lifted to access produce.

The National Wildlife Federation Program

The National Wildlife Federation (NWF) of the United States is a leading agency that furthers the cause of wildlife protection. An important way it does so is through its Backyard Wildlife Habitat Program. As early as 2003, more than 36,000 properties in the United States had been certified as participants. A Habitat Stewards Program, launched in 1999, trains volunteers to create wildlife habitats in their communities. A Schoolyard Habitats Program, offers workshops and educational materials and a Community Wildlife Habitats Project encourages residences and businesses to make "green space" for wildlife. For more than a decade a Campus Ecology Program has helped transform some of the nation's college campuses into living models of an ecologically sustainable society while training a new generation of environmental leaders.

It was in 1973 that the NWF leaders noted a striking statistic: 70 percent of Americans lived in cities and suburbs. Editor John Strohm wrote in the April/May 1973 issue of National Wildlife "fewer and fewer children will be able to rub shoulders with nature." How could kids "brought up on concrete" grow into adults who would care about wildlife? NWF had a plan: Bring the wildlife to the kids! So was born the idea—backyards, large or small, rural or urban—could, at little cost, be transformed into wildlife habitats, creating a network of "mini-refuges" for plants and animals displaced by development (See "When Gardeners Grow Wild," Berger, 2003).

"Landscaping with native plants is now a growing trend in the U.S., but in some places so are foreclosures for failing to follow rules that demand a manicured yard" (See Ridgley, 2005). Homeowners associations may make such demands, but fortunately for persons who wish to garden with native plants, attract wildlife, and often save on water bills, change is underway. Even some federal and corporate office buildings now make native landscaping a priority. Getting your property certified by the NWF may help.

If you contemplate getting involved, it is well to check in advance what the rules are if a homeowners association exists. One restriction may be related to fire hazard. It is of interest in this regard that native species are often more resistant to fire than non-native ones, sometimes not recognized by the rule makers.

If there is no restriction on natives, I recommend the focus be on "true natives," species that thrive in surrounding undeveloped natural areas. Don't go too far afield because native plant diversity and the native animals it supports may change over short distances. Check with local plant authorities—universities and botanical gardens and consult plant identification guides.

Should nature-oriented homes and corporate landscaping take hold on a large scale in the U.S. and abroad, the process could not only help slow the loss of much of our native plant life, but that of their many dependent animals as well.

LEARNING FROM SUCCESSFUL COOPERATIVE MODELS

CHAPTER FOUR

LEARNING FROM SUCCESSFUL COOPERATIVE MODELS

Since some schools lack even a modest representation of free-living animal and plant life, it is important that school districts consider giving high priority to the establishment of cooperative off-campus teaching and research areas for nature studies to which school children can be transported. Places may be found for such studies in established parks, on some reservoir watersheds, and on public lands leased or purchased for the purpose. Establishing such sites in rural areas will be much easier. These larger areas, with more complex ecosystems, can be an important adjunct to schoolyard nature study programs. The Barstow School District Nature Program was such an example. For five other outstanding models see the Appendix, p.209.

The Barstow School District Nature Program

This outstanding program was conceived and designed by Mr. Leon Hunter, the Districts' Science Specialist, and was in operation for 30 years from 1968 to 1998 in the 20 schools in the Barstow Unified School District in the Mojave Desert, California (Fig. 20). It is a model of what should be done, wherever possible in many different environments worldwide, to restore our connections with nature. Emphasis was on study of the local flora and fauna of a little disturbed section of the California Desert, conducted by the students and the teachers themselves. Mr. Hunter wrote the following history of the Desert Research Station (DRS).

The original project proposal was prepared in 1967. The application received federal funding under the Elementary and Secondary Education Act (ESEA) Title III, a funding source to encourage implementation of innovative ideas in all areas of education.

A 120 acre site of federal land was leased from the Bureau of Land Management in 1968 and a field laboratory was constructed that could serve a class size of up to 40 students. A nature trail and 10,000 sq. ft. pond was constructed. By the end of 1968 over 2000 students in grades 5 through 12 had participated in environmental field studies. The objective of the project was to involve students in

LEARNING FROM SUCCESSFUL COOPERATIVE MODELS

Fig. 20. Desert Research Station nestled among native vegetation of the Mojave Desert, Barstow, California, Secondary School District.

Problem-solving investigations choosing widely among more than 100 problems in desert ecology, animal behavior, solar energy, plant growth and plant physiology, soil and pond chemistry, and adaptations to the desert environment.

Before student visitation was possible, it was necessary for *on-site teacher in-service training where teachers engaged in similar problem solving.* (Italics mine) This was followed by student pre-visit preparation in which students would spend two weeks or more in selecting a problem, proposing hypotheses and working out methods for testing the favored hypothesis. This allowed the full day at the station for data gathering that could be evaluated back at the home school. Many problems required return trips with modified plans, or for further data gathering. Students at each grade level returned each year with new problems. In the lower grades (5-8) problems were handled as a class effort, with teams each working on certain aspects of data gathering. High school students worked in smaller groups or individually on problems of their choice.

A full time field instructor at the station worked to coordinate student activities, supply needed equipment and materials and help the regular science teacher with the class. *The field instructor was*

LEARNING FROM SUCCESSFUL COOPERATIVE MODELS

the key to success of the program (Italics mine). This position had to be filled by an enthusiastic instructor who was very knowledgeable about science methodology and desert ecology.

To provide for ever-changing needs and interests, additional grant proposals were written and these funds were used to provide animal enclosures for behavioral studies, construction of an animal room for temporary housing of desert animals, a small greenhouse for plant studies, solar energy equipment, computers, updates to the laboratory library, new in service workshops for teachers, and new science equipment and materials.

Since any in depth field investigation requires considerable time in the field, the station offered late afternoon classes for interested high school students. These students came out two afternoons per week throughout the year to work on individual investigations. This was in addition to the regular daily program for a class of students. Some of these students participated in the program during all four years of high school. The results of some of these student's work were so excellent that students regularly received awards at state and national science fairs, science talent searches, and science scholarship programs.

Some group environmental studies by high school students such as the student investigations of the Mojave Chub (an endangered fish) expanded over several years and students attended sessions of the Desert Fishes Council presenting slide talks on the results of their investigations concerning the monitoring of the fish population, growth rates, predation and fish behavior. More than 200 fish were tagged and using Lincoln-Peterson Index, students were able to keep track of the individual fish and postulate population numbers.

Over the years, returning students sometimes became so involved and interested in their field and lab investigations that they joined scientific organizations and environmental groups. Students learned to use university libraries such as the U.C. Riverside Science Library and how to do online searches to locate professional researchers who could help them with their scientific needs. All of these things took place during pre-college years.

DRS facilities not only served elementary, junior high, and senior high students, but also a number of college graduate students who needed field facilities for their desert research. Students from the University of California Los Angeles, U.C. Riverside, U.C. Berkeley, U.C. Santa Cruz, University of Southern California and the University of South Carolina used DRS facilities in connection with their investigations. Excellent cooperation by curators at the San Bernardino County Museum and the Los Angeles County Museum resulted in furthering the interest of Barstow students who sometimes assisted these visitors. The relationships that developed between high school students and visiting scientists sometimes resulted in a symbiotic sharing of work and information. Students were thrilled to have the opportunity to work with these professional investigators and were excited about serving as data gatherers and monitors of field recorders and weather instruments and taking animal census. Occasionally these relationships proved so valuable that the DRS students were paid a small salary from the grant money of the professionals. In several instances, the students' help was valuable enough to result in their names appearing in the acknowledgement section of the published scientific papers, a reward not easily forgotten by young, impressionable high school students.

In the 30-year tenure of the DRS's operation there was sufficient time for following the careers of some of the former participants. Although the DRS program's major objective was merely to help students become familiar with some environmental concerns and problems, as well as exposure to methods one might use to solve problems, a significant number of students became so interested in science that they majored in the various sciences in college, some completing masters and doctoral degrees. A number of others went into medical professions, but reported back to the DRS staff that this choice was greatly influenced by their exposure to science problem solving at the DRS.

LEARNING FROM SUCCESSFUL COOPERATIVE MODELS

It appears that a higher percentage of former students, than would normally be expected, became scientists or developed life-long hobbies dealing with science. Some became politically motivated to pursue environmental problems. As DRS students, some had attended Bureau of Land Management (BLM) advisory committee meetings and offered comments during the public comment periods. Others wrote letters to elected officials, joined environmental groups and science organizations. These activities suggest that their "hands on" field experiences with nature were not superficial and did not end in short-term enthusiasm.

As urban populations become ever more crowded, it would seem that the need for understanding the problems resulting with this overcrowding of the habitat, become ever more urgent. The ever-accelerating loss of plant and animal species through human activity indicates a desperate need for more awareness of nature by all people.

The biodiversity of this planet, which required 3 billion years to develop, must not be lost due to lack of man's basic knowledge of its existence. —Leon, Hunter, March 10, 2003.

Regrettably, the DRS no longer exists. With the loss of its leadership and failure to find teachers with adequate training and interest in ecology and the discovery approach to science, the program, as originally conceived, was discontinued. Students lost interest as emphasis shifted from a problem solving "let's find out approach" to more of a short term descriptive approach and lack of on-going follow-up procedures involved in research. Had we centers like the original Barstow program servicing schools all over the United States and abroad, we would be well on our way toward spreading the knowledge of the processes of science, the nature study goals, and the ecological literacy so important to our future (See Orr, 1992 and p.198—The Importance of Political and Administrative Support).

But the good news is the discovery "hands on" field approach to science conceived by Leon Hunter lives on—carried forward by his former students. Herewith, a recent letter of support for his program by one of his students:

Mark A. Wilson
Lewis M. and Marian Senter Nixon
Professor of the Natural Sciences
Department of Geology, The College of Wooster
Wooster, OH 44691 USA

Sept. 24, 2004

Dear Leon,

I wanted to tell you that I've been working the past two summer field seasons in the Negev Desert of Israel. My research (and that of my students and Israeli colleagues) is standard geology and paleontology, but I've also become involved in discussions of high school science education in the major town of the Negev: Mitzpe Ramon. Some friends of mine were struggling with how to take advantage of the expertise of visiting scientists for students (thereby improving greatly the learning), and I then regaled them with stories about our own Desert Research Station in the Mojave.

My Israeli friends have taken the DRS ideas very seriously, so we may soon set up the beginnings of a similar program for Israeli high school students. How's that feel, Leon, for having yet another time-delayed, long-distance influence on teaching?!

I hope you are doing very well. I often think of our desert adventures, and the foundations of my life in science (and in other inquiries), which you helped me build.

Best wishes,

Mark

LISTENING IN ON A NATURALIST'S EXPERIENCES

*Nature Walks. Field Trips.
Interacting with Animals in the Field.
Nature Stories.*

CHAPTER FIVE

LISTENING IN ON A NATURALIST'S EXPERIENCES

Field trips that focus primarily on wild nature are one of the quickest ways to start the nature-bonding process. I have watched it happen in conducting nature trips as a ranger-naturalist in the National Park Service, working with youth groups, and in my role as a zoology professor, conservationist, and father. With young people, a single outing may do the trick (See p.52)! Although it helps to be a well-informed leader, much depends on attitude—which should be "Let's find out." As parents we can use this approach with our children. Let's not let kids get preoccupied with computer games on a camping trip. There are far more important, interesting and exciting things going on in the real world of nature nearby if one takes the time to look.

Nature Walks and Activities

Having led many field trips and "nature walks," I have a few suggestions. With young children, if opportunity affords, take them on a short fast "hike" first, to let them "blow off steam" and get a little tired. They then will be more receptive to focusing on their surroundings. Also find things that will get them personally involved. Telling can be important but too much talking, then moving on, then more telling, can be boring.

Look for harmless subjects that can be handled and examined closely, on the spot—certain insects, lizards,* frogs and salamanders (but remember most have skin secretions that can "sting" the eyes so hands must be washed after handing and eyes not touched), wild flowers, leaves (their shapes and structures can be a cause for wonder). Look for the many ways plants are adapted to distribute their seeds. Search for animal "sign," tracks, owl pellets (open to look at bones, fur, etc.), scats (use a stick to break apart to try and tell what has been eaten), spiders and their webs (who can be first to spot a spider?). How do spiders differ from insects (find one of each, place in plastic or glass containers, and examine closely)? Catch, examine, and release butterflies (See p.148). Look for plant and animal interactions- insects and the plants they pollinate. Bring in caterpillars and their food plants to witness the marvel of transformation.

---

*Easy to catch and release unharmed (See p.148 and 149)

Useful tools are hand lenses and binoculars, dip nets for pond samples, and insect nets. Lizard nooses can be improvised on the spot (See p.144). Older participants with cameras can take pictures of subjects of interest.

Sit the participants down, in a place of at least partial concealment. Urge silence and limit mobility, and imitate an injured bird (See p.124). If things work well, you may have everybody wanting to learn how to do it.

Caution: Wherever you go *be sure to know wildlife protection rules* and the hazards—poisonous plants, venomous animals, etc. to avoid undue risks. Contact Fish and Game Departments, Park Naturalists, Universities, Botanical Gardens, Natural History Museums, and check Natural History Field Guides.

Herewith, details of a field trip I conducted in which I used a "hands-on" approach to get my group directly involved. Some may question my choice. But sometimes something dramatic is required to jump-start the wheels of curiosity about the natural world.

Woodcraft Ranger Guide Conference

September 30, 1944. Camp Ah-di-hi, NE of Mount Wilson in the Sierra Madre above Pasadena, California. (Based on my Field Notes)

An Antlion Story

There were 62 participants. Ranger Guides had the choice of a number of different events that were to occur the next day. At the evening campfire each leader described his or her program. Mine was "Nature Exploration." I told the guides, "I have found a sandy bank with many antlions traps. On our trip tomorrow, you could drop ants into the pits and watch the antlion attacks and even scoop up the sand under the camouflaged attacker, carefully sift through the sand, and see it!" (I thought, since kids like fantasies—the more gruesome and threatening the better, why not adults?). However, having seen the many interesting woodcraft activities, I wondered if anyone would show up for my nature walk. The next morning I was pleasantly surprised. Around 30 people were awaiting my arrival.

LISTENING IN ON A NATURALIST'S EXPERIENCES

Fig. 21A. Antlion pit-fall trap

The large turnout for "Nature Exploration" may have been due to the story I told about antlions at the campfire get-together, the night before. I used a ploy I learned during two summers as a Ranger Naturalist in Lassen Volcanic National Park, California. <u>Don't just tell people about nature but try to get them actively involved.</u>

The day before, I had checked out the area along a trail that led out of the campsite. I had not been in this part of the Sierra Madre before. It followed a shallow creek. I saw many Mountain Yellow-legged frogs (*Rana muscosa*)—adults, larvae, and metamorphosing individuals (a species that is now on the verge of extinction!). Other creatures seen were garter snakes, whirl-a-gig beetles, water striders, trout, and many birds. I noted the great diversity of plant life. There were lots of interesting subjects for folks looking for things to teach about "nature exploration." Most appealing for my purposes, however, was the sandy bank replete with antlion pit-fall traps and nearby ant nests. Here was my hands-on activity for my group.

Antlions (also called Doodlebugs—members of the insect family Myrmeleontidae) are small insects that ambush ants and other small prey that stumble into their funnel-like pit-fall traps in sandy areas. Like ants, they are widespread. (Fig. 21A)

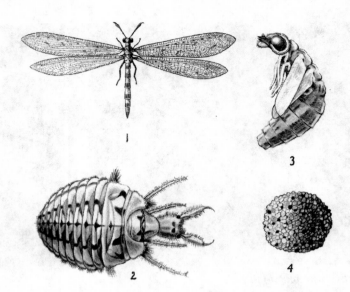

Fig. 21B. Antlion life stages: 1) Adult, 2). Larva, 3) Pupa, 4) Sand-covered cocoon

Adult antlions are delicate winged insects of the air that look like slender dragonflies. (Fig. 21B) The larvae are earth-bound in the pit of sand. The adult female lays her eggs in the sand, often in sheltered locations. The camouflaged larvae that emerge may grow to around 10 mm or so in length, are colored like the sand, have a flat head and long scimitar-like hollow jaws with which they pierce their prey and suck its blood. The larva makes its trap by throwing out sand with an upward jerk of its shovel-like head. When the trap is finished, the "lion" lies buried at the bottom of the pit and extends its jaws in readiness to grab its prey. The sides of the pit are as steep as the sand will lie so when an ant or other small insect falls in, and attempts to climb out, the sand cascades downward. The "lion" may use its head-shovel to toss sand at the struggling prey to keep it from climbing out of the trap and to bring it into range of its jaws.

Activity: Antlion Pitfall Trap

Place an antlion in a container with slick vertical walls (large coffee can?) with fine sand from the collection site. The depth of the sand should be around 2-3 inches. With patience you may see the antlion make its pit-fall trap. Sometimes it may occur at night, however. Feed your captive aphids, ants, small beetles, or other small insects dropped into the pit.

After awhile, the "lion" will pupate in a hollow "ball" of sand lined with self-produced "silk," beneath the bottom of its pit, and if you are lucky, after some weeks, you may see the adult (imago) emerge, cast off its pupal covering, climb up a nearby object and slowly unfold and spread its delicate wings. Place a stick with a flat surface near the location of the pupa that extends high enough (3" or so) to allow the wings to expand (See also Note 5, P.1, Another group of insects that builds funnel-shaped sand pit-fall traps).

A Clockweed Story

(On internet, check <u>Fabulous Filaree</u> for more information). To give the plant world equal billing, I follow with an activity pertaining to the way a highly successful plant, Clock Weed, has adapted to planting its seeds—of equal interest in getting people involved outdoors. (Fig. 22)

On nature walks I often look for plants of special interest in getting people focused on nature (See also Stokes and Stokes, 1985). Among my favorites is Clock Weed (also called Storksbill or Filaree), genus *Erodium* in the Geranium family *(Geraniaceae)*. *Erodium* occurs in Europe, Eurasia, Australia, North Africa, and temperate America, thus is widely available for study. They are annual and/or perennial herbaceous plants that usually occur in dry, open or disturbed sites. It is the seed and its associated structures that especially attract attention.

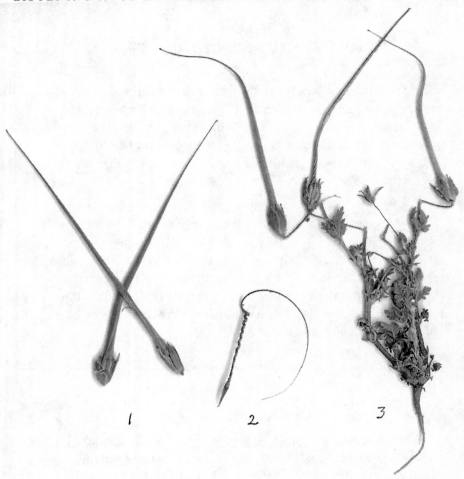

Fig. 22. Long-beaked Storksbill (Erodium botrys) Native to S. Europe; widespread in California) showing (1) "scissors" made from two pods; (2) released fruit with sharp-tipped seed attached to the columnar coiled carpel with its' slender curving style; and long tap root.

As a child someone (probably my Mom) showed me how to make "scissors" from a pair of still green Clock Weed seed pods (probably the species *E. botrys* with large pods) done by making a slit with my thumbnail toward the base of one pod and sliding the other through the opening. (See Fig. 22) But there are things much more interesting and important to do as follows:

LISTENING IN ON A NATURALIST'S EXPERIENCES

Look for fruits around the base of the plant. A Clock Weed "fruit" consists of a "seed," a sharp-pointed, slender, tapered cone, about 3-7 mm long (depending on the species) attached to a "carpel," an elongate corkscrew-like columnar structure and a "style" that flares out at its tip into a slender curving scimitar-like shape. Botanically speaking, the seed with its attached paraphernalia is a fruit just as a peach with its pit is a fruit.

Clock Weed fruits may lie dormant in the often-dry areas where they occur until there is a rise in humidity or actual rainfall. Then a remarkable event begins. As the fruit absorbs moisture, the corkscrew carpel begins to uncoil and its attached seed tip and scimitar-shaped style move with it. Slender, resilient, whitish hair–like structures that project from the sides of the carpel (be careful not to damage them), increasing in length toward the style, tip the carpel and its attached seed at around a 30° angle, even on a flat surface. This aims the sharp-tipped seed toward the ground where its slightly curved tip (see with hand lens) may catch on cracks, insect holes, or other surface irregularities. The rotating tip promotes a burrowing action of the seed. In the often-baked dry environments where most species occur, soil cracks are frequent and seeds often take hold in the cracks, each sending down a long taproot.

Since there is usually considerable dead leaf and stem debris (these are annual or biennial plants) and other impediments around the base of Clock Weed plants, the long arching movement of the style is likely soon to be impeded or stopped. The result is transfer of additional energy to the uncoiling carpel and seed. Watch the seed closely with a hand lens to see its slow burrowing action.

Look at the surroundings of the plant. You may find fruits in varied positions—some lying flat, others at various angles, some standing upright. Perhaps much depends on support of surrounding objects that influence the angle of seed penetration. However, you may find few, if any fruits with seeds, if an ant nest is nearby. Ants harvest the seeds.

Once a seed takes hold in the soil its numerous short bristles, pointing away from the tip (see with a hand lens), support penetration and would tend to resist removal (collectively acting like an imbedded arrow point). There is a weak place between the seed and the carpel where, at some point, the carpel and the style break away from the seed.

Since the carpel responds to fluctuations in humidity and rainfall by coiling and uncoiling, it appears that once a seed has gotten a start, say in a crevice, it would rotate repeatedly in one direction then in the other, digging ever deeper in response to humidity changes until soil friction stops the action and the carpel and style break away.

Clock Weed has several methods of dispersal. Fruits are light and, as noted, are provided with elongate hairs. Put a fruit on a flat surface and provide a puff of air to see how responsive it is—easily moved by air currents. Fruits also are readily caught up in animal hair and in clothing. In *Erodium cicutarium*, a species with relatively small fruits I have studied, runners that radiate out from the plant base along the surface of the ground, sometimes reach three feet or more in length, greatly increasing the plant's reach.

Why called Clock Weed? Thoroughly moisten the fruit and hold the seed so that the style faces you. Look directly at the top of the carpel. Watch the style. Its movement resembles a clock hand. Moistening the fruit by mouth with saliva, however, is not advised because of the risk, especially with children, of swallowing it and getting it lodged in the throat or trachea. Instead, apply saliva with your finger to the carpel or, better, dip the fruit in water or spray it with an atomizer.

Activity: Clock Weeds in Action

(1) Scatter Clock Weed fruits on a tray of soil rubble scraped up from the habitat and scattered lightly with dead leaves, stems, etc. Often many fruits are found around the base of a plant. Wet the assemblage with water from an atomizer. Watch the fruits spring into action as the styles begin to move with the uncurling of the carpels. Look for fruits whose seeds have "found" a toehold. Select one and watch the seed closely with a 10X hand lens to see its slow rotation. Note direction of movement, which will correspond to that of the style.

Since movement of the seed may be very slow, depending on fruit size and humidity changes, patience is required to see the action. If you watch long enough you may witness the reversal of coiling direction as the fruit dries.

Look closely at and near such sites to see the results of seed planting. At one site I found a fruit with its seed completely buried in dry hard soil, and the carpel, with style in place, as erect as if lined up with a plumb bob. At another site I found seed and carpel out of sight in what appeared to be a small vertical insect hole with the style attached and lying flat on the ground surface.

(2) Because Clock Weed fruits are so sensitive to changes in humidity you can create a small hygrometer. Glue the fruit in an upright position to a thick piece of flat white or pale cardboard upon which you can mark with a dot (use a pencil) the location of the tip of the style at intervals when viewed <u>directly from above</u>. To ensure that the carpel is vertical tease it into position in the drop of glue as it is drying. It may help to cut off the tip of the seed before placing it in the glue. When the glue dries you are ready to record humidity changes.

Compare readings taken in the living room with those taken shortly after a shower in the bathroom. Time the intervals and connect the dots. If the style reverses direction, use an arrow to record the change and trace along side the previous record. If you wish detail, keep track of time between dots. Record changes at a single location from time to time throughout the day. What happens at nightfall?

Recording air temperature may also be of interest in relation to rates of change. Wet a Clock Weed fruit removed from a refrigerator and compare rate of uncoiling with the same fruit wet at room temperature.

Other plants have fruits that respond to moisture and temperature changes like Clock Weed: Wild Oat (*Avena fatua*) and Purple Needle-grass (*Nassella pulchra*), but their structure is somewhat less complicated. Try wetting them to see the fruits in action. You may enjoy experimenting with them.

LISTENING IN ON A NATURALIST'S EXPERIENCES

Combining Works of Nature and Man

In addition to focusing on nature in the wild it is also important to make some field trips comprehensive, including the activities of man and the historical factors that have affected the land and its use.

In the 1960s the Walnut Creek Elementary School District in Contra Costa Co., Calif., embarked upon a program of field trips that emphasized such breadth. Joanne Taylor, a teacher in the district, concerned with the importance of environmental education, was the intellectual leader (See Taylor, 1963). She was assisted by U.C. Berkeley Botanist, Dr. Mary Bowerman.

Classes of fifth and sixth graders were transported by bus on overnight trips to a variety of places in the San Francisco Bay area, noted for their historic and natural features, and to one site, Norden, in the Sierra Nevada. Many trips were taken over a period of 15 years.

One of the outstanding features of the fieldwork was the combination of involving children with "hands-on" activities in the field, followed by their written accounts of their experiences. Note the many topics and student authors in the following sample report, reproduced as prepared by the students.

Of special interest were "Plot Studies" (underlined in above report). These were geometric quadrants marked out with string on a ground surface for recording features of physical and/or biological interest. A grid of nine one-foot-squares was created. Students, seated on the ground, sketched the location of plants, animal sign and other features of interest using colored pencils, and took care to locate items accurately within the grid. The activity promoted attention to detail and hopefully a greater respect for undamaged natural ground surfaces and some of nature's smaller features. Help with species identification was provided by adult overseers.

The report was illustrated with many drawings (some in color) and there were several black and white photographs of participants taken by the students. Taylor emphasizes that the <u>attitude</u> of the leader is <u>crucial</u> to the success of field programs, for if the leader lacks enthusiasm, this will "rub off" on the students.

Temalpa
Mount Tamalpais Study, April 2-3, 1963

Introduction
    A. Log of the trip   by Shari* and Susan
    B. Maps
        1. Marin County   by Jay
        2. Topographic Map   by James
        3. Trail Map   by Susan
        4. Trails of Mt. Tamalpais   by Laura
        5. Bootjack   by John and René
    C. Radar on Mt. Tamalpais   by Kathi
    D. Climate Graphs
    Temperature for Mt. Tamalpa   by Julia
    2. Temperature for Kentfield   by Julia

II. History
    Indians
        1. Indians of Marin   by Laury
        2. Daily Life   by Jean
    B. Legend of Mt. Tamalpais   by Sherry

III. Recreational Areas
    National Monument
        1. John Muir   by Shari
        2. National Monuments
        3. Muir Woods National Monument pamphlet
    Other Recreational Areas

IV. Coast Transition Life Zone
    Grassland—Chaparral Habitats
        1. Grassland—Chaparral habitat
        2. Serpentine
            a. <u>Plot Study</u>   by Lesley and Susan
        3. Plants
    Mushrooms   by Richard

---

*Last names deleted to provide anonymity.

<u>Plot Study</u>   by Jean
    4. Animals
       a. Insects   by James
       b. Scorpions   by Jean
       c. Reptiles   by Kurt
Mammals
    1) Deer   by Gail
    2) Blacktail Deer   by Carol
    3) Mice   by Judy
    4) Rabbits   by Judy
Forest Habitat
    1. Redwood—Canyon Habitat
    2. Plants
Trees
    1) Ditto
    2) Coast Redwood   by Jim
    3) Trees   by Mark
<u>Plot Studies</u>
    1) Oak-Fir   by Gail and Kathy
    2) Base of trees   by Judy
    3) Stream   by Kurt
Canyon
    1) Ditto
    2) Oxalis   by Lesley
    3) Ferns   by Roger
    4) Patterns
    3. Animals
       a. Amphibians   by Steve
       b. Forest Birds   by Judith
       c. Mammals
    1) Raccoons   by Judy
    2) Bobcat
       a) Report   by Carol
       b) Prints   by Michael

V. Bibliography

LISTENING IN ON A NATURALIST'S EXPERIENCES

Humans and the Importance of Soil

It was on a trip to visit our daughter Mary in Vernon, British Columbia, Canada. She was teaching elementary school in nearby Lavington, and was devoted to getting her students interested in nature. It seems we had hardly finished unpacking, when she told me she had arranged a nature field trip and I was to lead it! My students were to be teachers from her school. When I asked what was to be the objective, she said it was for me to decide. "I'm not going to tell you what to do."

That night I slept fitfully worrying about what lay ahead. What should be the focus on what might be a one-time opportunity to perhaps influence a group of teachers in whose charge were youngsters at an impressionable age? The topic, I felt, must be of great importance. Around 3:00 AM I had the answer. It must be the importance of soil.

My childhood on a farm, and later my efforts as a scientist, along with other scientists, trying to stem the tide of indiscriminate soil-damaging off-road vehicle (ORV) activity in the California Desert, surely influenced my choice. The ORVs were a long hard, on site, lesson for me from 1973 to 1980!

On a previous trip to BC I had seen a road cut that could dramatically portray the importance of soil—a steep, perhaps 20ft. slope of pale, glacial till, at the top of which was a thin dark layer only about 2ft. thick. On its edge stood a dense, forest, surely weighing many tons. The message must be that without this dark fertile layer, here and throughout the world, our planet's terrestrial life and perhaps even much of that of the oceans would be as lifeless as the moon—no plants, no animals, no people! So we spent the day in the forest seeing how soil is formed.*

---

*I learned recently from my daughter that several of the teachers on that trip of many years ago were "turned on" to develop soil programs for their classes. One of them prepared an ongoing unit of study on soil in which writing and reading skills were woven into the science of soil study. At this writing, 2008, the project continues.

Fig. 23. A section of little-disturbed California Desert creosote bush habitat before (left) and after (below) a single ORV race involving approximately 700 motorcycles.

It is a complicated and lengthy process involving plants, animals, microorganisms, vulcanism, breakdown of rocks, the atmosphere, and the actions of weather over vast stretches of time—measured in centuries and millennia. An excellent example of the importance of an animal in soil formation is the Pocket Gopher *(Thomomys monticola)* that occupies the California Sierra Nevada. The digging activities of this animal, present in great numbers, acting over vast stretches of time, have been a major factor in creating the soil fertility of the Great Valley of California—an agricultural wonderland.

The Pocket Gopher, armed with their heavy incisor teeth and sharp claws that remove soil, can even break away softer rocks. Their mixing of vegetable matter with soil particles as a result of their herbivorous diet, have greatly contributed to the rich soil that has been carried by stream action to the Valley floor.

Soil damage or loss to human activities can thus be serious and life-threatening and recovery time extremely slow.

Therefore, study of soil, its life-giving properties and its care should be high priority in the education of everyone—especially now in a machine age driven by rapid population growth and environmental ignorance of, or disinterest in the soil, that is capable of causing such severe damage.

Particularly fragile soils occur in deserts and tundra but all soils must be treated with great respect and care. Some U.S. television car advertisements don't help!

The Off-road Vehicle Onslaught. *(Figs. 23–26)*

Of special interest to me has been the fragile biotic (living) wind-resistant soil crusts found widely throughout deserts. Once destroyed the surface, in dry weather, can "bleed" interminably with dust. One dirt bike race can create damage that will last for decades. The millions of these machines and their "relatives" (See below) now operating in many parts of the world, and widely in the U.S., are creating severe, and often irreversible damage to soil and biota, including the spread of weeds that out-compete native species and, in some areas, can lead to an increase in desert fires. Impacts are particularly damaging in arid lands because of the fragile surfaces, exceeding long recovery rates, the sheer number of these machines, and the difficulty of maintaining regulatory control (See Darlington, 2003). It is therefore very important that riders stay in designated ORV parks or other areas legally open for their use and that the use there be managed so that there are not continual requests for new areas as the land becomes gullied and dusty. Land is too precious to be so treated.

Fig. 24. (Above) My former graduate student, Ron Marlow, blowing on the area of undisturbed desert crust (no dust plume). (Below) In the same general area nearby, blowing on the surface pulverized by passage of the dirt bike race. Loss of desert crust ensures chronic dust problems because of slow recovery.

Fig. 25 Shadow Hills area, Eastern Mojave Desert, California, showing heavy tracking by ORV motorcycles. The scars are expected to last for many decades. Photo by H. Wilshire.

Fig. 24A. Close up of desert crust before race.   *Fig. 24B. Close up of desert crust after race.*

Consider the following: A typical ORV motorcycle (Dirt Bike), with its knobby tires driven in a straight line impacts around one acre in traveling around 20 miles; typical 4-wheel vehicles disturb one acre in about 6-8 miles of travel; and 3-wheel "all terrain vehicles (ATVs)" disturb one acre in about 4-6 miles of travel (the lower soil compression effect of the balloon tires unfortunately is offset by higher shear stress which is very disruptive of vegetation and

Fig. 26. Dark clumps of cryptogamic crust. Navajo National Monument, northeastern Arizona. Photo by Steve Abbors.

surface soil structure. (See Webb and Wilshire, 1983, Wilshire p.629 in Latting and Rowlands, eds., 1995, and Wheat, 1999)

Developing reverence for the soil—a lesson in antiquity.

On trips to the California deserts with my students, I often talked to them about <u>biotic crusts</u>, prevalent in deserts. I would dig a shallow hole a foot or so wide so they could see, in profile, this protective vital surface layer, often only an inch or less thick. Each person could then see rootlets and the tiny filaments of fungi exposed on the crust's undersurface—the mycorrhizae that attach to plant roots and are so important in undisturbed environments in aiding plants to obtain nutrients and water and are nourished by the plants in turn.

Biotic crusts are often referred to as "cryptobiotic ("containing concealed life"). They are "a tightly knit assemblage of lichens, fungi, algae, mosses, and cyanobacteria*.

---

*Cyanobacteria (also called blue-green algae) go back to the beginning of life on Earth and were the primordial producers of oxygen (followed by an expanding diversity of plants) that prepared the atmosphere for the flourishing of life. These are tough organisms, even having found a way to live under extreme desert conditions. I have found colonies on the underside of quartz rocks in the California Desert. Look for white translucent rocks. Pick up a partly imbedded rock. Look on the underside. You may see bluish or yellowish green patches (the colonies). Hold the rock toward the sun to see how light passes through, allowing the colony's chlorophyll to function. For conformation check opaque rocks. Colonies are not expected. Check other deserts and locations for light-transmitting (diaphanous) rocks.

Each of the crust's constituent organisms holds little biological sway on its own. But collectively, they do nothing less than to hold together the desert floor" (Royte, 2003). Although extremely fragile to mechanical impacts, they are enduring over vast stretches of time in natural undisturbed environments. A study by Jayne Belnap of the U. S. Geological Survey found that such crusts in the arid Utah desert can take thousands of years to reach maturity (See also Faden, 2004). Thus heavy damage by off-road vehicles or other surface disrupting activities can mean chronic dust storms and a severe impact to the desert biota, including its human inhabitants. "Crypto," as it's affectionately called by rangers in Arches and Canyon Lands National Parks, has been reported as making up approximately 70-80 percent of all living ground cover in the Colorado Plateau (See McCluskey, 2004).

Cryptobiotic crusts are widespread. In addition to fostering plant growth, and protecting the ground surface, they are a critical component in global climate, especially in pale-surfaced arid lands. Although pale early in development, when mature, their usual dark color (often black) decreases reflectance of sunlight. This results in less warm air rising from the earth's surface, that may drive away cooling clouds, increase the atmosphere's temperature and reduce local rainfall.

"Some scientists liken the urgency of protecting desert plant communities—and the soils that sustain them—to the urgency of saving rainforests". (See Bainbridge and Virginia, Desert soils and biota, in Latting and Rowlands, 1995 and Royte, 2003, Don't spoil the soil).

Pertinent are the words of two leaders speaking a half-century apart—a Native American and a South African environmentalist:

Chief Luther Standing Bear speaking of the Lakota, the western bands of plains people now known as the Sioux:

"The Lakota was a true naturalist—a lover of nature. He loved the earth and all things of the earth, the attachment growing with age. The old people came literally to love the soil and they sat on or reclined on the ground with a feeling of being close to a mothering power. It was good for the skin to touch the earth and the old people liked to remove their moccasins and walk with bare feet on the sacred earth. Their tipis were built upon the earth and their altars were made of earth. The birds that flew in the air came to rest upon the earth and it was the final abiding place of all things that lived and grew. The soil was soothing, strengthening, cleaning, and healing."

Alan Paton, 1948, in "Cry the Beloved Country" (South Africa):

"The grass is rich and matted, you cannot see the soil. It holds the rain and the mist, and they seep into the ground, feeding the streams in every kloof. It is well-tended, and not too many cattle feed upon it; not too many fires burn it, laying bare the soil, Stand unshod upon it, for the ground is holy being even as it came from the Creator. Keep it, guard it, care for it, for it keeps men, guards men, cares for men. Destroy it and man is destroyed."

Field Trips of Good Will

Karl P. Schmidt, dean of American Herpetology, believed that naturalists would be ideal for the Foreign Service. He pointed out that they would seldom get bored during assignments abroad, even in austere environments, because of their love of nature. Therefore, they could be counted on to stick-it-out under adverse conditions and their interest would rub off on the local people and garner their trust. A naturalist is soon followed by a bevy of children, then by adults. It is not difficult to trust a person who is occupied with collecting plants, carrying an insect net, bird watching, or catching lizards!

My former graduate student and a family friend since his childhood, Dr. Ted Papenfuss, is such a naturalist who has traveled to many parts of the world—China, Russia, the Middle East, Pakistan, Vietnam, Africa, Mexico, and South America as he carries out field work on vertebrates for U.C. Berkeley's Museum of Vertebrate Zoology. Was it some of his early youthful experiences on trips afield in California that set the stage for his adult commitment? It was through his efforts that the Chinese now have a herpetological society and respected scientific journal, Asiatic Herpetological Research. Wherever he has gone, he has been an ambassador of good will and has made many friends, not only among field biologists but persons in many other walks of life.

When Ted was 11 years old he joined a Boy Scout troop along with my son, John. I would go on some of the weekend Scout trips and teach the boys about natural history and show them how to identify different species of amphibians and reptiles. Ted never lost his interest in herpetology. He works in countries that are often considered difficult for Americans.

He first went to China in 1986 to meet China's leading herpetologist, Ermi Zhao. During this visit, Professor Zhao suggested that they start a collaboration to study the reptiles of the deserts of China. Two years later, Ted went to Leningrad, USSR to plan a collaboration with Soviet scientists on the reptiles of the deserts of the USSR. There he met with the leading herpetologist in the

USSR, Ilya Darevsky. In 1991 he helped organize a joint visit to the University of California for both Zhao and Darevsky.

In 1995 he attended a meeting in Turkmenistan of the Asiatic Herpetological Research Society. There he met Dr. Cao Von Sung, the Director of the Institute of Ecology and Biological Research (IEBR) in Hanoi, Vietnam. Dr. Sung had been a general in the North Vietnam Army during the Vietnam War. One evening Dr. Sung came to Ted's hotel room with a gift of a small jade turtle. He told Ted that in Vietnam's culture, a jade turtle was a sign of friendship. He said it was time to forget the mistakes of 20 years ago and start to work together as friends. Dr. Sung invited Ted to come to Vietnam in 1996 as a member of a joint Russian-Vietnamese expedition to study the amphibians and reptiles of Tam Dao National Park.

After the 1996 trip, Ted helped write a grant to continue the research and in 1997 a University of California-Russian Academy of Science-IEBR returned to Tam Dao to complete the survey. The scientists spent a month in the National Park surveying the amphibians, reptiles, and small mammals. The National Geographic Society arranged for a writer and photographer to join the expedition. An article, Tam Dao-Sanctuary Under Siege, was published in the June, 1999 issue of the National Geographic Magazine.

Ted went to Iran in 1998 and started working with Iranian biologists. A year later he organized an American-Russian-Iranian expedition. This was the first American scientific group to be allowed to travel freely throughout Iran since the 1979 Islamic Revolution. They even went to the Afghanistan border in search of endemic species.

Some of the mainly Afghan species were not found and Ted decided he needed to go to Afghanistan to complete the research. At the time Afghanistan was controlled by the Taliban, a group not known for its love of Americans.

Ted found a telephone number for the United Nations Ambassador from the Islamic Emirate of Afghanistan, the name the ruling Taliban had given the country. He talked to the First Secretary, a Mr. Zadran, telling him of his interest in visiting Afghanistan to collect lizards. After a long pause, Mr. Zadran, said "Yes, I think that is the type of visit that we can help you with."

LISTENING IN ON A NATURALIST'S EXPERIENCES

During the fall of 1999 and spring 2000 arrangements were made for Ted to travel to Kandahar, Afghanistan. He flew to Quetta, Pakistan where he was met by two Taliban representatives who had traveled overland from Kandahar to Quetta. Ted traveled with the two men to Kandahar and was put up in a guest house in Kandahar. The Taliban Corps Commander for Kandahar, Mohammad Akhtar Usmani, visited Ted at the guest house. He was very interested in Ted's research and Ted showed him some lizards that he had caught in the garden. An English-speaking warrior-guide named Mohammad and two senior Taliban warriors traveled with Ted on his journey around Afghanistan. He was also given a travel pass that allowed him to visit any place in Afghanistan. Ted and the Taliban soon became good friends and all enjoyed traveling about looking for lizards.

One morning, in the hills south of Kabul, Ted tried unsuccessfully to noose large agamid lizards. Mohammad was watching Ted and came up to him and said "Let me try." He handed Ted his Kalishnikov assault weapon and took Ted's lizard noose, a fishing pole with a slip noose tied to the end. Ted waited at the car. In a few minutes a smiling

Fig. 27. Afghan guide, Mohammed, returns with his quarry. He had become a proficient lizard catcher.

Fig. 28. Native people of Sierra Leone, Africa, some holding gift flashlights (some still in boxes) to be used in hunting for their rediscovered frog. Dr. Ted Papenfuss stands on the right.

Fig. 29. *Cardioglossa aureoli*, a tiny frog, (around 3/4 inch snout to vent length) revered by the native people of Sierra Leone, recently rediscovered, and now being considered as the national emblem.

Mohammed returned with a large agamid lizard dangling at the end of the noose. For the rest of the trip, the Kalishnikov stayed in the car while Mohammed used the noose.

In July 2004 Ted went to Sierra Leone in Africa to obtain some frogs for DNA study. Sierra Leone had suffered from a six year long civil war that ended in 2004. Most of the infrastructure had been destroyed. Ted worked with a biologist named Abdulai Barri from the local university who wanted to find a tiny species of frog known as *Cardioglossa aureoli*. The frog had been discovered in 1963 and had not been seen since. One night while Ted and Abdulai were looking for frogs in the Western Area Forest near the capital, Freetown, Abdulai rediscovered the *Cardioglossa*. (Fig. 29) The discovery made national and international news and the

LISTENING IN ON A NATURALIST'S EXPERIENCES

government of Sierra Leone promoted this tiny frog as a symbol of survival after the terrible civil war. Even the American Embassy issued a press release on the discovery.

There are many people, including young and old, working throughout the world now on our environmental and social problems. Many started with a childhood love of nature, perhaps triggered by a family or school field trip. We must continue to nurture such experiences.

Interacting with Animals in the Field

Imitating Animal Sounds

Learning to imitate animal sounds can contribute significantly to nature-bonding and the enjoyment and understanding of animal behavior, and it is easy to learn to do so. Can you imagine a greater thrill in getting acquainted with wildlife than to have a wild animal come close in response to your imitation of its call! It can add greatly to the enjoyment of field trips. Owls are a good subject with which to start. They are widespread and most are easy to imitate.

Owl-Calling.—Professor Loye ("Padre") Miller, ornithologist/paleontologist, and chairman of the zoology department at UCLA, taught me how to imitate the Great Horned Owl (*Bubo virginianus*) and my family, and I, on field trips, have had the pleasure of "playing" with Horned Owls for many years. Herewith, are two episodes based on my field notes. (Fig. 30)

April 11, 1941. Colorado River, vicinity of Big Maria Mts., Riverside Co., California. We were camped on the bank of the Colorado River, about 10 miles NE of Blythe. We had constructed a make-shift tent in case of rain. My companions, my brother Ernie and friend Esther, slept in the tent, while Anna-rose (Esther's sister) and I slept outside, nearby, on the ground. (Fig. 31)

Fig. 30. Great horned owls, female on right with ear tufts (plumicorns) depressed.

About 3:00AM I heard two male Great Horned Owls ("talking" to each other). One sounded NE, up river; the other across the river. I gave a few sleepy hoots and went to sleep.

I was awakened a little later by a light rain so we moved into the tent. Heard an owl again. Began answering. In a few minutes heard a rustling of branches perhaps 50 yards up river. Hooted again. The branches rustled once more and to my delight the owl appeared, flying silently and, directly toward me, ghost-like in the moonlight, and landed on the ridge pole of our tent, not over 5 1/2 feet above my head! He hooted, I hooted. He peered over the edge of the tent with ear tufts (plumicorns) extended, moving his head from side to side as if to more carefully check the interior for "the other owl."

We hooted back and forth. Then a scratching sound indicated he had taken flight, followed by a thump as he landed on the ground nearby. Moonlight again illuminated the area. He walked with stately tread toward the tent coming within 10 feet and appeared ready for battle (characteristic bowing pose with tail elevated each time he

Fig. 31. Field notebook sketch made after the Great Horned Owl episode on the bank of the Colorado River, north of Blyth, California. Note aggressive pose of owl as it approached our tent. Upper Left: Presumed location of rival owl across the river.

hooted). I asked Anna-rose for the flashlight, but our talking and movement frightened him. He flew up river. Shortly thereafter, we again went through a similar pattern, the owl finally alighting on the tent and once more landing on the ground, this time in back of our tent.

I got ready with the light and partly covered my head with the sleeping bag top and began hooting. Suddenly he appeared on the ground at the corner of the tent within 3 feet of my head. I was so surprised I reflexly turned on the light and lost the chance to see if he would have actually attacked. I had planned to duck into my sleeping bag. Perhaps it is just as well he departed. The talons of a Great Horned Owl are formidable indeed. He remained in the vicinity until 8:00 AM the following morning.

In the meantime, he moved around our tent from one perch to another, several times alighting on top of our parked car. Finally he perched in a cottonwood tree about 100 feet away and continued hooting. When I got up to explored the area for other wildlife, he followed me as I periodically mustered a feeble hoot (hooter wearing out!) to keep him interested.

The super aggressiveness of this owl, I believe, relates to the other male I heard across the river. The river was probably a territorial boundary that had been trespassed, and at a time when the young would be demanding special parental care.

August 25, 1941. Summit Lake, 6700 foot elevation, Lassen Volcanic National Park, California.

As a Ranger Naturalist in Lassen Park, I was expected to give talks and lead songs at evening campfires for the general public. One of my talks was on owls, in the course of which I imitated the birds that occurred at Lassen Park. Carl Swartzlow, a big 6 foot Swede, the Park Naturalist, a geologist by training, was in charge of the duties of the young naturalists. Occasionally he sat in on the programs conducted by his newly appointed tenderfoots.

One evening, after hearing my owl talk, he said, "Can you really call owls?" He obviously was skeptical. I assured him I could and I was soon to prove it.

An afternoon outing and dinner for naturalists and their wives and children took place at Summit Lake, at the cabin site of a fellow naturalist. Swartzlow was present. The children, playing outside, reported seeing a large owl fly to the top of a pine tree. Carl and I, accompanied by others, went outside to a clearing in the forest north of the cabin near where the children had seen the owl. I began hooting. The bird moved and we all could see him at the top of a pine north of the clearing in the fading light of sunset. It was, indeed, a Great Horned Owl. I continued hooting but the bird was silent. Then I got a Horned Owl response, northeast of the clearing. By the increasing volume of its call it was evident that it was approaching—this in spite of all the versions of Great Horned Owl given by children who had gathered in the clearing. The first owl remained silent.

All at once, things began to happen. The owl that we had first seen, but still not heard, left his perch and flew silently, silhouetted against a twilight sky, to the top of a lodgepole pine some 50 feet from us at the edge of the clearing. Almost simultaneously the second owl appeared at the top of a pine about 25 feet to the east, 75 feet or so from the first owl. Both owls were now hooting and so were the children. The birds didn't seem to mind.

Owl #2, upon spotting #1, promptly flew toward him and bumped him off his perch. #1 left the area evidently having transgressed a territorial boundary and was not seen again.

Owl #2 now began flying from one tree to another across the clearing as we exchanged hoots. It was getting dark but there was moonlight. All but one youngster had tired of the show and had gone to the cabin. Carl, one of the older children, and I remained. Now just one "owl" was hooting from the ground.

The owl began swooping lower and lower as he flew across the clearing. He also appeared to be triangulating on my calls as he shifted his traverse routes. I had noted this on previous owl-calling episodes. Triangulation seems to be their method of pin-pointing the location of an intruder. The three of us were standing like statues to not frighten the bird. At last, he swooped toward us and poised in the air directly above Carl. With wings flapping rapidly, yet silently, he lowered his legs and spread his talons about to land on Carl's bald head, which may have been glistening in the moonlight. All three of us "hit the dirt." From now on we held our coats over our heads.

The bird was tenacious in its efforts to deal with the intruder. It landed at the top of a 20-foot lodgepole pine nearby. Carl approached, yet the bird did not leave until he shook the tree. Even after this disturbance he remained in the area. An hour later, after the dinner party was over, and guests were leaving, he was still tenaciously hooting his territorial challenge. Carl was now a believer!

Activity: Imitating a Great Horned Owl

It is easy to imitate a Great Horned Owl. See the cadence below. If you hear one, practice matching cadence, pitch, and sound quality. The pitch will probably be at your lowest. As the footnote shows, the pitch is usually uniform. My daughter Mary, who has a higher-pitched voice than I, has had success. She has found by lowering her jaw and rounding up her lips she is able to closely approximate the bird's sound quality; so females can also participate. Don't slur the separate sounds. Each should be abrupt and given in staccato fashion with a mellow falsetto quality. A straight vocal sound may not do the trick. The call of the male is most likely to yield results.

**Great Horned Owl cadence - - — — or - - - — —
Calls are about 4 or 5 seconds in length.**

Hide and Seek with Owl Calls

A person who can imitate a Great Horned Owl hides in an area of wild growth, in a well-concealed location. The "owl" should avoid wearing bright clothing and lie low (Watch out for poison oak or poison ivy). He gives the call of the male at intervals of 10-15 seconds (approximating a cadence I have often observed). The calls should be separated at times by pauses of 20 seconds or more. Group members hunt for the owl. Since the call of a Great Horned Owl has a ventriloquial quality, finding the caller is not as easy as one might think. The person who finds the owl gets to be the next one to hide. The winner is the "owl" most difficult to find. Triangulation will help (See p.125).

LISTENING IN ON A NATURALIST'S EXPERIENCES

Great Horned Owls are widely distributed from the tree line in North America to Tierra del Fuego, at the tip of South America, and from coast to coast. They occur in varied habitats—forests, woodlands, open country and deserts.

Many of my encounters have begun in late afternoon and evening. After you learn the call of the male, you can sometimes start a discourse even if you haven't heard an owl. Best results are likely to occur when the birds are nesting and caring for young. Nesting in some areas can begin in January. Failure to get a response may mean you are outside the territorial boundary, or you may need to work on your owl call.

The Eurasian Eagle Owl (*Bubo bubo*)—of Eurasia and North Africa is the Old World counterpart of the Great Horned Owl. The two species are very similar so probably the Eurasian Owl would respond similarly.

The importance of persistence, and refining ones imitation on the spot, in owl-calling was made clear to me on a trip to observe the eastern Barred Owl (*Strix varia*) near Vernon, British Columbia. Our naturalist guide, Jim Grant, said the birds were showing up in B.C., having dispersed west across the country. It was a cold dark night as my daughter and I followed him to the site in a dense forest. Upon arrival we stood quietly listening. Finally, Mr. Grant began an imitation, whistling low notes into his cupped clasped hands in a cadence and sound that seemed to say, "who cooks for you"—"who cooks for you-all." After some 10 minutes of fruitless efforts, I suggested I might try. I had never heard a Barred Owl, but Loye Miller had taught me the call as he knew it from the southeastern U.S. where the country folks say the bird says "Who'll cook for my folks if I cook for you-all?," perhaps a somewhat different dialect. Jim guided me on the technique until he decided I had a "good" imitation.

Barred Owl cadence with a falsetto -- -- -- -- all on one pitch but dropping at end—hoohoo—hoohoo—hoohoo—hoohooaw. This represents a typical call. At times a great variety of sounds are heard.

However, I used the Loye Miller cadence, a <u>falsetto</u>, and a somewhat higher pitch than a Great Horned Owl. After 10-15 minutes, with no response, we decided to quit. As we began our hike out of the area, I continued to hoot but now had lowered the pitch. Suddenly there was a single prolonged series of rapid hoots that started high and descended in pitch. I immediately copied the sound and the owl responded. After some vocal exchanges, he suddenly appeared—landing on top of a broken pine stub within 20 feet of us, looked down at us for a few moments in the beam of my flashlight, then flew off to join another bird, presumably his mate, where a "conversation" ensued. As we left the area, I continued to hoot and the owl followed until we were presumably out of his territory.

Playing Parent to Baby Horned Larks
(*Eremophila alpestrist*)

Barstow Unified School District Desert Research Station, Vicinity of Hinkley, Kern Co., California, April 22, 1980 (Partly from my Field Notes)

I came upon a nest with very young Horned Larks. In 1972, in the California Desert, I had found that whistled sounds at a certain pitch range could cause nestling young Horned Larks to respond as though the parent had arrived. The song of this bird has been described as a "weak tinkling or twittering." The young are hatched naked and with eyes closed, thus at this stage, they must depend on sound and tactile information to detect the arrival of a parent with food. Nests are built on the ground, often in the shelter of a bush or grass tuft.

I decided to retest my whistled pattern and pitch on the naked and blind Hinkley young birds. My procedure in my 1972 trials was to position my mouth about one foot above the babies and to use a cadence, all on one pitch, starting at my lowest whistle, pausing, then moving up the scale, one note at a time. I would then repeat the sequence in reverse (going down the scale). The cadence used in 1972, purely by accident as I found later, was that of the opening of Beethoven's Symphony No.5, C minor, – – – —! but all on one pitch and greatly muted.

Fig. 32. (Left) Young Horned Larks, still blind, respond to my whistle. Note black dots on their tongues said to aid the adult in targeting their mouths with food. (Below) Young bird recently off the nest wearing camouflage plumage during flight training.

As before, at almost two octaves above my lowest whistle and at higher pitches, the little birds extended their necks and opened their mouths. The babies made me think of blooming flowers as their orange mouth linings rimmed with yellow were suddenly displayed and as they swayed their heads rapidly from side to side, their mouths widely open in anticipation of food.

The Horned Lark is a bird of open, often sparsely treed ground, and is widespread throughout North America. However, being a ground nester, it is highly vulnerable to ground surface disturbances. Territorial males fly to great heights, singing and displaying with fluttering wings. They then dive, resembling small dark bombs, often involving several pullouts, to eventually alight on the ground near the nest site. In so doing they declare their "ownership" of a column of air and the ground below. The famous Skylark (*Alauda arvensis*) of Europe is an ecological counterpart. Might its nestlings also respond to human-whistled sounds?

An Amphibian Response

My success in attracting animals through mimicking their call has not been confined to birds. I was making my way on foot at night toward a distant amphibian chorus that started at the end of a thunderstorm in the desert in southeastern Arizona. I was aware that at least three species were calling—the Great Plains Toad (*Bufo cognatus*), Green Toad (*Bufo debilis*), and Couch's Spadefoot Toad (*Scaphiopus couchii*) and that I had joined many silent, shadowy, hopping forms headed in their direction, presumably mostly females. In the light of my headlamp, I spotted a large Couch's Spadefoot and decided to try and divert her by imitating the voice of the male—a plaintive nasal cry somewhat resembling the anxious bleat of a lamb. We had been moving together toward the chorus perhaps 30 feet apart. The animal seemed oblivious to the light. As I began an imitation she stopped and, to my delight, turned and abruptly hopped directly toward me, coming within a yard or so of my feet!

Field study of birds and other animals can be greatly aided by learning their songs, calls and other sounds. A person with such knowledge can determine presence and location of many species without seeing them and distributional and census studies can be greatly accelerated. Time spent learning animal sounds, therefore, will not only provide the excitement and enjoyment of attracting wild animals, but it will greatly improve one's effectiveness as a field observer. Sounds can be timed, in some cases including their intervals; compared to familiar sounds; described, including means used in attempts at imitation; and recorded graphically, showing changing volume and pitch, by a change in thickness and a rise and fall in a line; and cadence by breaks in a line. Sometimes mechanical or other devices can be used. I have gotten the Northern Cricket Frog (*Acris crepitans*) to respond by imitating its metallic <u>gick, gick, gick</u> call by striking two smooth stones together with a sliding motion, and the Western Chorus Frog (*Pseudacris triseriata*) by stroking the small teeth of a pocket comb.

I don't feel I have any special talent for mimicking animal sounds. I believe it is a matter of incentive, taking the time to listen carefully, and perhaps a willingness to feel foolish as one experiments. The voices of some species are, of course, simply too complex to imitate, but there are many that can be mimicked quite easily. Once you have a wild animal respond to your imitation, you may become addicted. However, be careful with large raptors. Loye Miller was struck on the forehead by a Spotted Owl (*Strix occidentalis*) in southern California when imitating the owl's call.

Use of Distress Sounds

Many animals are attracted by the distress sounds of their own or other species. Learning to imitate the generalized distress sound of birds can often be useful in bringing a concealed bird or other animal into the open where it can be photographed and/or identified.

I first witnessed the effectiveness of this imitation on the UCLA campus in Los Angeles. I was in a natural history class taught by Professor Raymond Cowles. He was introducing us to birds of the campus. He had paused at the crest of a slope, below which was the gymnasium, perhaps 50 yards away, where the lawn nearby was dotted with black forms—Brewer's Blackbirds *(Euphagus cyanocephalus)*. Cowles put his closed fist to his mouth and began making a loud squeaking sound interspersed with agonizing squeals. The class stood spell-bound as the entire flock of perhaps 50 or so birds became airborne, wheeled about as though triangulating on the sound, then headed directly toward us. The birds "rained" down about us alighting on bushes and the ground, some within a few feet, as we stood motionless watching the amazing spectacle.

To make the sound, liberally moisten the curled index finger of your clenched fist with saliva and produce a repeated kissing sound by placing the side of your mouth firmly over the base of the finger while allowing a little air to leak in on the other side. Draw out some of the sounds to increase their agonizing quality. The technique seems to work almost anywhere in the world, and in obtaining a response from birds, is usually most successful during the breeding season.

Distress calls can also attract predators. I once called in a coyote I had glimpsed at a distance of perhaps 50 yards. I alerted my partner, a motion picture producer and photographer, to ready his camera as we hid downwind behind a bush. The animal responded quickly and came running to within about 30 feet before it saw us, made a quick turn, checking us out as it did so, and sped away. The picture became the opening shot of a nature film.

LISTENING IN ON A NATURALIST'S EXPERIENCES

Locating Animals by Triangulation

Locating a small calling animal hidden in a large expanse of rough terrain, especially at night, would appear to be almost impossible. However, it can usually be accomplished easily by means of triangulation. The technique is particularly helpful for finding creatures whose voices are ventriloquial and thus give a deceptive impression of location. Triangulation is best done by two people. When, for example, a calling frog has been singled out of a chorus and its approximate position determined, move 15 to 30 feet (5-10 m) apart and listen quietly. After a few moments, each person, without discussion, should decide on the location of the sound. Then, at a signal each should point with arm extended and sight on a distant object beyond the expected location of the sound that will serve as a reference point. Finally, both should walk directly toward their reference points and seek the animal where their pathways cross. If you are alone, you can listen at one position for a time, decide on a direction, then move to one side and listen again. Triangulation is often easier to do at night than in the daytime if flashlights or headlamps with distinct beams are used; the point where the beams intersect can be determined precisely. The lights should not be turned on, however, until direction has been determined. Sudden illumination may alarm wary species, and they may not call again for some time. First trials may not bring success, and it may be necessary to withdraw and repeat the procedure several times.

I've often used triangulation in locating amphibian choruses in arid lands of the American southwest. After thunderstorms the response of frogs and toads, long ensconced below dry sun baked earth, may be dramatic. Within an hour an area powder-dry for many months may reverberate with their cries, and the ground may swarm with hopping forms. However, breeding sites are often spotty and a distant chorus may require triangulation to locate. When heard far away on a night drive, triangulation by car may be required at the start. However, great care must be exercised to stop well off the road.

Small scale triangulation can be used at night to locate small creatures such as stridulating katydids, tree crickets, etc. Work in pairs or step from side to side. It may be necessary to keep your light off until you are quite sure of the location of your quarry.

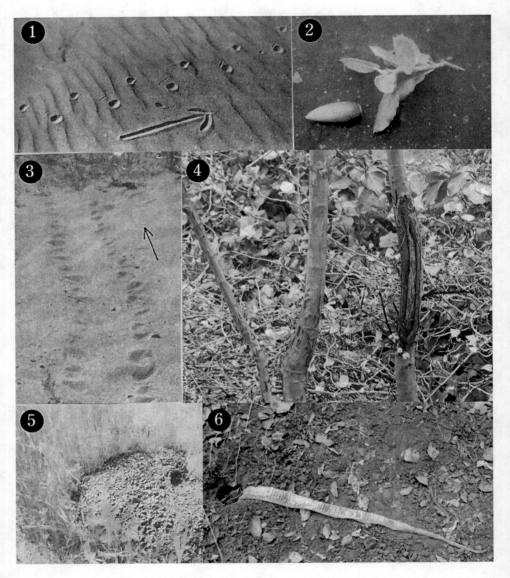

Fig. 33.
1. *Fringe-toed Lizard track—sand dune near Palm Springs, California*
2. *Coast Live Oak (Quercus agrifolia) branchlet broken off by a scrub jay (Aphelocoma californica) to obtain an acorn—a routine procedure.*
3. *Desert tortoise track—Pinto Basin, Joshua Tree National Park, California*
4. *Sapling scraped by a Black-tailed Deer to remove velvet from antlers.*
5. *Pocket-gopher (thomomys bottae) tailings. Note burrow opening, rarely left open.*
6. *Skin shed by a Gopher Snake (Pituophis catenifer) as it exited a gopher burrow.*

Tracking and Animal Sign

Seek areas of fine, loose soil, sand, or mud—sandy flats, dunes, dusty roads, trails, fresh mud of washes, or the sandy banks of ponds and streams. Go out when the sun is low and highlights and shadows are strong. If possible, walk toward the sun. Tracks then will be seen in relief. Start early, before there is a maze of tracks, or later in the day after a wind has erased old tracks and new ones are appearing. Follow a fresh track. Direction of travel can be determined by toe marks and ridges formed by the backward pressure of toes, feet, or coils of a snake. Tracking demands paying attention to details and using clues. From meager evidence an interesting story may unfold.

When I was working on my doctoral thesis (early 1940's) in the Garnet sand dunes near the then small town of Palm Springs, California, I tracked a lizard across the barren rippled surface of a dune. The track indicated that at first the animal had moved slowly. Marks of all four feet showed, the stride was short, the tail dragged. Then there appeared the tracks of a Greater Roadrunner (*Geococcyx californianus*), a lizard-eating bird. The lizard's stride suddenly lengthened and marks of only two feet could be seen; the tail mark disappeared. The lizard was running now, on its hind legs with tail lifted. An occasional small dent indicated that at high speed it occasionally touched down with a front foot to maintain balance. Just over the crest of the dune the track suddenly stopped. To one side was a faint V-shaped mark in the sand. The roadrunner track continued at full clip over the hill, then slowed and wandered. The bird seemed confused. I grabbed at the V-shaped mark and something wriggled beneath it. In my hand I held a beautiful Coachella Valley Fringe-toed Lizard (*Uma inornata*). Its skin of minute round scales feels like velvet in one's grasp. (Fig. 34)

In the smooth mud of a wash bottom, near Palm Springs, I followed the track of a lizard. To one side a Sidewinder (*Crotalus cerastes*) track appeared. The Sidewinder is a venomous snake of the American Southwest. The tracks converged and became entangled. A fragment of lizard tail, probably cast during the struggle, was found, but only the sidewinder track continued. (Fig. 35)

Fig. 34. Courting Coachella Valley Fringe-toed Lizards (Uma inornata). Female (in foreground). Scene painted before windmills—snow-capped Mount San Jacinto in the distance.

On another occasion, in the same wash, the tracks of a coyote and sidewinder met. The place was a maze of contorted snake tracks and coyote prints, but only footprints left the area!

Nature-bonding involves not only gaining knowledge about the pleasurable aspects of nature but also risks. A sidewinder leaves a distinctive track when sidewinding, the locomotion it uses when on a sandy or slippery surface, in its usual habitat. (Fig. 35) In mid-April 1954, a party of U.C. Berkeley professional entomologists, Ray Smith, John MacSwain, and Gorton Linsley, their wives, and six children encamped on a sandy slope in Short Canyon, about 10 miles northwest of Inyo-Kern, Kern Co., California. They did so despite the fact they had seen sidewinder tracks in the area, but since they were naturalists, I can understand that they would not have been put off by what they saw because sidewinder tracks are common in the California Desert and a single snake in its night-time travels, can make many

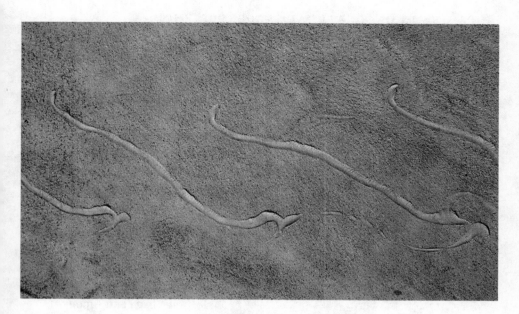

Fig. 35. (Above) J-shaped tracks left by a sidewinder. The hook of the J, at the top of the picture was made by the snakes neck. The hook points in the direction of travel. As the track is completed there is often a streak left by the trail (bottom of track). (Below) Sidewinder crawling. Note sections of body not touching the ground.

LISTENING IN ON A NATURALIST'S EXPERIENCES

Fig. 36. Sidewinder (Crotalus cerastes), an arid land snake of the North American Southwest and Northwestern Mexico. Highly venomous.

tracks. At bedtime they slept in their sleeping bags, at least two members of the party (perhaps all?), on the ground. However, this party was in for a surprise. Several female sidewinders exuding sex pheromones had entered the area.

At 5:00 AM Dr. Smith heard a sidewinder sound off within 2-3 feet of his sleeping bag and upon arising found a sidewinder track within 2 feet of his sleeping wife's head. At 5:30 AM, Mrs. Linsley had almost stepped on a female snake coiled in a shallow pit in the open where its color blended closely with background. That evening a small male had rattled close to the group's campfire.

A morning survey revealed sidewinder tracks "everywhere" converging on the campsite, and the next day Dr. Linsley found a pair side-by-side.

Dr. MacSwain brought a gallon jar crammed with five live adult sidewinders, 3 males and 2 females, to my laboratory, all caught at the campsite. Two pairs had earlier copulated in the confines of the jar—such was the power of their sex drive! The entomology group had unknowingly positioned itself in a sex arena triggered by at least two female snakes that had entered the area, followed by their ophidian suitors.

LISTENING IN ON A NATURALIST'S EXPERIENCES

Desert Tortoise Stories

In the California Desert I have many times found Desert Tortoises (*Gopherus agassizii*) by following their distinctive tracks—an animal of football or larger size, but sometimes difficult to find when it is resting quietly in rocky terrain, where they resemble a rock, or when sheltering in concealed locations under bushes. Often their track may lead to burrows where they may spend over 90% of their time. (Fig. 37)

Desert Tortoise Research Natural Area—6½ miles NE California City, Kern Co., California

I had heard reports that the Desert Tortoise was "fond of people" and would approach a quiet person to "check him or her out," yet I had not had such an experience.

October 25, 1975 (Based on my Field Notes).

On site at the above tortoise research area where my graduate student, Ron Marlow, was engaged in field studies of tortoises. 10:45 AM. Spotted an adult tortoise basking at the opening of her burrow-at about 30 yards. Decided to see if she became alarmed at my approach by stopping at intervals and counting her eye blinks (assuming that increasing frequency would suggest increasing anxiety???). Started out lying flat on ground then, at later stops, sat upright with notebook on ground. 11:05 AM. I'm now within 12 yards. Tortoise still at mouth of burrow. Rate of eye blinks has steadily increased. Decided to come back later for further testing. Carefully retreated.

2:10 PM. Cautiously approached tortoise. She is resting with limbs sprawled, well away from her burrow. Eyeblinks now fastest so far (maybe because air temperature highest yet and air drier). Engaged in careful stalking to within 15 yards, after she became preoccupied with feeding (2:18 PM). She had "licked her chops" just before starting to forage. Was this in anticipation of food? I sat motionless. After eating grass for a while, she rested. Then to my delight, she abruptly got up and crawled steadily down slope toward me, stopping in my cast shadow where she again rested. There were plenty of bush shadows nearby. Why had she stopped at mine? Was it my red-colored pen? She then again smacked her "lips" and moved off to forage about 25 feet away (2:32 PM).

Fig. 37 (Above) A tortoise can be mistaken for a rock. (Below) Tortoise burrow at Fort Irvin, California, occupied by a young female especially attractive to the males in the area. When she tires of male attention, she retreats to her burrow, escaping the persistent attention of the large dominant male, too large to enter her burrow.

Fig. 38 The tortoise approached me, as I crouched motionless, from a distance of about 10 feet. It is sniffing my shoe.

The next day, October 26. 11:45 AM.

Tortoise abroad. I carefully approached her and lay down within 3 feet, with my notebook open nearby. After about 5 minutes she crawled toward me and placed her gular horn (a prong extending beneath her throat from the underside of her shell) on my notebook! She seemed completely unafraid. After several minutes, she crawled past my open notebook touching it with tip of her nose and my outstretched body, tapping my pant-leg with her nose. She then headed upslope, stopping at around 25 ft. and began again eating grass (11:55 AM). Marlow confirms that the animal was a "wild" tortoise that had not been in captivity.

A recent experience adds to the observed repetitive nature of this interesting behavior.

March 26, 2005 (From my Field Notes). Base of south slope of Pinto Basin Dune, Joshua Tree National Park, Riverside Co., California.

My two daughters, Melinda and Mary, and I, encountered a young male Tortoise abroad about 3:00 PM. Temperature mild; mostly clear; many wildflowers in bloom. As we stood quietly by, at a distance of about 10 feet, he showed no sign of fear, and did not attempt to move away. During the entire episode, lasting perhaps 20 minutes, he never withdrew his head, a common protective response. (Fig. 39)

Fig. 39 Young male Desert Tortoise. Vicinity of Pinto Basin Dune, Joshua Tree National Park, California.

At my suggestion, Melinda, where she stood, assumed a crouching position. The tortoise crawled towards her and hunkered down between her legs. She got up and moved a short distance away, again crouching. The tortoise repeated its behavior. Daughter Mary assumed a crouched position. The behavior was repeated. Twice I tried with the same result (Fig. 38). Was the tortoise seeking shelter or perhaps shade? My former graduate student, Ron Marlow, whose PhD dissertation was based on several years of field study of the Desert Tortoise believes they are very curious animals, responding to novelty in their environment. If we had been coyotes, the behavior would probably have been different.

Tracking involves determining direction and recency of travel, species identification, and sometimes interactions. The more thoroughly one knows the fauna and flora of an area the more useful and interesting the activity becomes in understanding unseen events. Determining what may have happened often involves detective work. When tracking, we rekindle the excitement of the hunt, a mainstay of human existence.

Nighttime Observing: Searching for Eyeshines

Fortunately for the nighttime observer, many animals' eyes reflect light. One of the pleasures in the field is to walk quietly through wild country at night in search of eyeshines, pausing occasionally to illuminate the surroundings with a headlamp or flashlight. Dewdrops and the eyes of spiders may glint silver and green, those of moths and toads often yellow or red; a murky stream becomes a cascade of light. To obtain an eyeshine, the light source must be held near your eyes. Hold a flashlight at the side of your head or with the base resting on your forehead; a headlamp works well and frees the hands. The eyeshine method works best on nocturnal birds, mammals, toads, frogs, and turtles. Eyeshines of snakes (except very large ones) and salamanders are usually too faint to be seen well, and, aside from geckos, most lizards are chiefly active in the daytime.

"Collecting" with a Camera

Many people now have digital cameras that provide beautiful colored photographs. With a home printing dock or compatible home color scanner, prints can be made promptly and placed in a family file creating a record of outings, and a life record of wildlife species identified. A wonderful opportunity is thus available to collect pictures of wild plant and animal subjects and their habitats on field trips. Get the children involved!

Photographing plants and their parts—leaves, flowers, fruits, is easy, but animals are much more difficult. Herewith, some guidance concerning birds, lizards, and insects which make excellent subjects.

Activity: Photographing Birds

Birds are often difficult to photograph because of their mobility and because they tend to keep at a distance. Here are a few suggestions:

1. Establish a well illuminated bird feeder near a window at your school or home.

2. Ready your camera, on telephoto if necessary; then use the distress call procedure described on p.124. Birds will often come into close range or emerge from cover nearby, alerted by the sound. Some may come from distant locations so give them time to arrive. Calling from a somewhat concealed location helps. Avoid quick movements.

3. Take advantage of feeding and nesting behavior, when they are preoccupied. Watch for approaches to nesting sites when they are feeding and caring for young (but don't get too close), and when they are engaged in mobbing behavior directed at predators. At such times many birds of varying species and sizes may assemble and vocalize, milling and darting about in efforts at harassing a common enemy. If you hear such sounds, approach cautiously. The object of attention may be as small as a Northern Saw-whet Owl (*Aegolius acadicus*), a domestic cat, or as large as a fox or coyote.

4. Birds are early risers. Vocalizations often start before sunrise and activity can be great in early morning hours, so start early. On a cool morning, when basking, some species may permit quite close approach.*

5. During the breeding season, males are strongly territorial and sing from conspicuous locations they habitually use. They may move from one familiar singing post to another. Such regularity provides opportunities for the photographer. The territorial song, as beautiful as it may be, has been referred to by ornithologists as a delightful form of profanity—"keep the h*** out of my territory!"

---

*On one occasion, when sitting quietly in dense brush, I had a Bewick's Wren (*Thryomanes bewickii*) briefly alight on my knee.

LISTENING IN ON A NATURALIST'S EXPERIENCES

Getting Close to Lizards

(For our purposes focus is on diurnal species)

Lizards are widely distributed and abundant in many areas across the planet, even ranging to high elevations in the Andes, the California Sierra, and elsewhere, so there are many opportunities for the photographer. Furthermore, many are easily captured for temporary and close-up observations and can contribute greatly to nature-bonding. Although some can be caught by hand, most must be noosed, a process that provides a sensation like that of catching a fish without the trauma to the animal fish-catching often entails.

With a little guidance (See p.142–143), noosing allows people to take in hand a wild, often beautiful, creature that can usually be caught and released without harm to either captive or captor (Fig. 40A). Only a few species—the Gila Monster, the Mexican Beaded Lizard, and Komodo Dragon, including a few of its close relatives, are known to have venom glands. However, lizards can bite. Although small ones, two to two and one half inches in snout-to-vent length, may merely pinch the skin, bites of some larger ones call for particular care in handling so follow instructions.

Fig. 40A. Collard Lizard (Crotophytus collaris), a splendid subject for a lizard nooser or photographer.

Fig. 40B. Lizard noosers in action.

To find lizards, go out early when they are in the basking phase. At that time many will allow close approach. Horned lizards *(Phrynosoma)*, found in western and central America, make great subjects because when not fully warm they are especially inclined to sit tight (Fig. 41). If you approach cautiously you can often get an unposed picture in a natural setting. This behavior, and their remarkable camouflage, may make them hard to find. However, watch for them on berms and rocks as you drive slowly along desert roads and search in the vicinity of ant nests. Start fairly soon after sunrise because they arise early in order to catch ants before heat shuts down ant activity. The Thorny Devil *(Molock horridus)*, a delightful, although very spiny Australian counterpart of horned lizards (with an unfortunate name), is also an ant feeder abroad when ants are active. Other species may be more difficult to approach, but a good stalker can usually get close.

You may not need to go far to find lizards if you are in suitable habitat. Many can be found at or near the roadside. Over many years my friend Nathan Cohen and I traveled together to obtain reptiles (including lizards) for classroom display and study, a practice now constrained by conservation concerns.

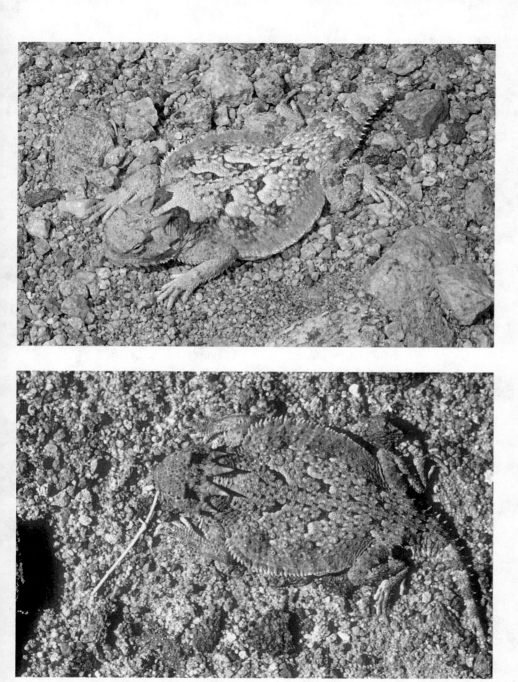

Fig. 41. Desert Horned Lizards (Phrynosoma platyrhinos), lizards that often "sit tight" when approached, apparently depending on their concealing coloration and form for protection. Northern Mojave Desert, California.

LISTENING IN ON A NATURALIST'S EXPERIENCES

Cohen's Law

Nate was stout, therefore he preferred to search near our parked vehicle, while I took off across the countryside. Over the years he seemed to do better than I in finding reptiles. It finally dawned on us why. Along many roadways pull-off sites to park are limited and thus repeatedly used. Human refuse, garbage, etc. tends to accumulate. This attracts insects and other creatures, food for lizards and rodents, and they, in turn, attract snakes and other predators.

A trip along California US Highway 8, near the Mexican border, made clear the "refuse effect." At that time, there was little human habitation except for a Stuckey's restaurant. It was hot so we stopped. But I was so energized for the hunt I took off across the desert. Nate chose to search around the restaurant. After an hour or so, I returned, red-faced, tired, and empty handed to find Nate relaxed and sipping a soft drink in the cool of the restaurant. I could tell by the look on his face, a teasing grin, that he had, once again, up-staged me! He reached into his collecting sack and brought out a Western Banded Gecko *(Coleonyx variegatus)* and a Western Shovel-nosed Snake *(Chionactis occipitalis)*, found under trash around the restaurant.

These experiences led to the formulation of "Cohen's Law: Success in finding reptiles and amphibians is inversely proportional to the distance from one's vehicle!"* The message is—don't overlook the chance to find these animals, including lizards, in the vicinity of your parked car, especially if a garbage container is present or there are other signs of food scraps. (Fig. 42)

---

* The accompanying cartoons by the author were given to Nate during his two bouts of illness and are shown here in his memory. Nate was one of my students and a dear friend and companion. He was chairperson of the Science Department at Modesto Junior College and later became Director of Science Education at U.C. Berkeley Extension. He was coordinator for the Extension's International Scientific Expedition to the Galapagos Islands in 1964 that I attended.

Fig. 42. Cohen's Law: "Success in finding reptiles and amphibians is inversely proportional to the distance from one's vehicle." Below. There can also be other problems in finding wildlife.

Activity: Lizard Noose Construction and Use

Temporary Capture

Remember to follow wildlife protection rules. For wary species capture may be required to get a satisfactory photograph. Tools and techniques follow.

When warm, most lizards are too fast to catch. A slip noose of strong thread, string, fishing leader, copper wire, or even a wild oat stalk, can be used to snare them (see below). The noose should be tied to a notch at the end of a slender stick, 3 or 4 feet long. A trimmed willow branch or other slender branch or stalk can be used, and often improvised on the spot. The noose length usually should not be more than 6 in. when the noose is open. If excessively long, it may become tangled in vegetation or be blown about so that it is hard to control. A wire noose avoids this difficulty and can be bent to thrust into small openings. Make a small loop about 1/4-inch in diameter at the end of the noose strand. Tie the loop with a square knot so that it will not close. Pass the shank through the loop and attach it to the pole. With a string or thread noose, should it tend to close when in use, open to the desired diameter and moisten both loop and shank with saliva where they come into contact.

To Make a Wild Oat Noose

Pull up a green stalk, being careful not to bend it to the breaking point. Strip off all leaves and the seed-bearing branchlets at the tip. Bend the slender tip around the stalk and form a small loop through which the stalk is to slide. Secure the small loop against closure as shown in Fig. 43, No. 3.

To Make a Copper Wire Noose

Cut a 10 or 12 inch length from an electric light cord. "Zipcord," obtainable in most hardware stores, can be used. Separate the two parts of the cord. From one of them remove the insulation and separate out a single strand of wire. Set aside the others for later use. To do so twist the ends of the bundle from which a single strand has been removed in opposite directions so that they will not separate.

Then form the bundle into a loose, open, overhand knot for convenience in carrying. A reserve supply may be needed (Fig. 43, No. 5 and 6). Now twist a small loop of 1/8–to 1/4 inch in diameter at the end of the single strand. Pass the shank through and orient the loop so that the shank moves freely. Compress the sides of the loop slightly to make it somewhat elongate; then form the open noose to a rounded shape to conform to the lizard's neck. In attaching the noose, take several turns around the end of the pole and twist the free end of the wire along the shank to strengthen its base. It is here that most breakage occurs.

Before using the noose make sure that all directions have been followed so no breakage will occur and leave the escaped lizard with a noose around its neck! After noosing a lizard, reduce the diameter of the noose loop to 1/2 inch or less and <u>carefully untwist all kinks</u>. Pass the shank between thumb and index finger to straighten it. Re-form the noose to the desired diameter and restore its rounded shape.

Noosing a Lizard

In noosing a lizard, avoid quick movements. When the noose is within 5 or 6 inch of the head, move it slowly, or pause for a moment, allowing the animal to become accustomed to the presence of a strange object nearby; then move the remaining distance gradually. When the noose has passed over the lizard's head and has reached the neck region, jerk upward and slightly backward. Remove the animal quickly, before it has a chance to wriggle free. Wary species can sometimes be noosed by creating a diversion. Gently shake a handkerchief at arm's length to one side or wriggle your fingers to attract attention away from the noose. Sometimes a lizard may jump at the noose, responding to its movement as prey. If this happens, reexamine the noose to be sure it is open for another try.

Removal of the Noose

To minimize trauma to the lizard and to avoid being bitten in removing the noose, hold the lizard in your left hand (if right handed) with your thumb on its back, above and between its forelimbs, and index finger against its throat. Hold its left forelimb between your index and second finger. If the lizard struggles to escape apply pressure on the forelimb to increase your grip. My hold is far enough back from the head to expose the noose for removal by my right hand. With a large lizard with a long neck, don't get too far back because the animal may be able to turn its head and bite. The surest way to avoid traumatizing the captive is to sever the noose with a scissors and make a new one.

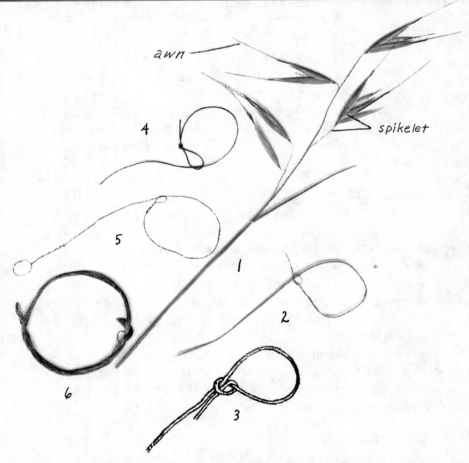

Fig. 43. (1) Wild oat fruiting stalk. Each spikelet usually contains at least two seeds, indicated by the projecting awns. (2) Lizard noose made from the tip of the fruiting stalk, after stripping away the spikelets. (3) Shows how to tie it. (4) Thread noose made by creating a loop with an overhand knot and passing the shank through it. (5) Noose made from a single strand of copper wire from a bundle removed from zipcord. (6) Zipcord backup bundle of reserve strands coiled for ease in carrying and handling.

Activity: Photographing the Captive

Equipment

(1) One or more gallon jars with screw top metal lids drill-perforated with 1/8 inch holes, for containing captives (Do not use lids with punched holes because of raw edges that may cause injuries). (2) Ice chest. (3) Plastic bags (Ziplock).

A reminder—in handling captive reptiles for study and enjoyment, as with any other wildlife, their well-being must be high priority. Thus when embarking on trips afield to observe wildlife, it is important to abide by legal measures to protect them. National parks and Monuments are off limits without special permits and state and federally protected species must be known and in no way harassed. The following instructions thus apply only to unlisted species, unless you have obtained a permit to capture listed ones.

Procedure

1. Place the captive in a glass jar or plastic ziplock bag with a crumpled paper towel to provide cover. I usually use a gallon glass container with a lid with drilled holes to provide air exchange.
2. Note and flag the site of capture.
3. Place the container with its contents in an ice chest to immobilize the animal*.
4. While cooling is occurring return with the ice chest and captive to, or near, the site of capture and ready the setting and camera focus using a stick or other object of appropriate size to simulate the subject. Consider lighting and background.

When all is ready, test the subjects' level of mobility. If hardly able to move, place it quickly in position for photographing because small lizards can warm up rapidly.

---

* Cold anesthesia is a harmless technique for temporarily immobilizing reptiles (and amphibians) as long as prolonged freezing temperatures are not employed. Indeed, brief exposure to cold numbing-temperature is well-tolerated by most "cold-blooded" ecothermal vertebrates (as contrasted with the "warm-blooded endothermal" birds and mammals). Cold anesthesia is even used in some surgical procedures in humans!

Take pictures as the subject warms up. Watch the eyes closely if you wish to avoid a photo of a sleepy-looking lizard. When warming, lizards often close their eyes. Since you have returned your captive to the place where it was caught, it will be familiar with the area as it escapes. An animal outside its home range is at high risk.

Activity: Using Tonic Immobility to Quiet Lizards

When you are ready to release a lizard, before doing so, try demonstrating "tonic immobility." (Fig. 44) If you are right handed hold it by the sides of its head in your left hand, back against your palm. Then place your right hand, fingers together over the animal, and slowly rotate your enclosing hands so the lizard is in a horizontal and upright position. Allow a brief pause. Then with the animal still enclosed, quickly reverse position so that it experiences being suddenly turned upside down. The "throwing on the back" movement is what tends to set off tonic immobility, a harmless transitory state. Still holding the lizard by its head, open your remaining fingers to expose the rest of the animal. Then with your right hand gently stroke the lizard from its chest region toward the rear several times while holding your left hand in a steady position. Now slowly open your left hand and release the hold on the head. The lizard will usually lie quietly, but for deep breathing, for a minute or so before making its escape. The procedure gives you a chance to have a look at the underside of the animal. If you wish a longer look or wish to take a photograph, repeat the procedure before the animal is aroused. Many other creatures—chickens and other birds, amphibians, sharks, etc. engage in tonic immobility. Perhaps it is often a form of "playing dead" causing a predator, especially an inexperienced one, to become careless in handling its prey.

Fig. 44. Tonic immobility (sometimes referred to as "playing dead"), here revealing the colorful ventral coloration of a breeding male Western Fence Lizard (Sceloporous occidentalis). In fighting or mating displays these colors, often not seen, are suddenly revealed by fattening the sides.

For the Insect Enthusiast

(Take into Account Protected Species)

Many insects can be easily stalked and directly photographed and there is a remarkable diversity of subjects. You can truly engage in "collecting with the camera" for these many species. However flying insects, of which there area also many, will usually require the use of an insect net. In what follows, I describe how to make a net (should one not be found in a nature shop). I also provide guidance on how to go from net to photograph, hopefully without trauma to the captive, which can then be released unharmed.

Making an Insect Net

Use organdy for the net, a yard of heavy wire (but bendable) for the frame, a discarded broom handle (or one-inch diameter dowel) for the handle. The handle should be about 3 feet long. The net should be bag-like, about 2 feet long, somewhat tapered toward the tip and with a circular opening at the top, about 10 inches in diameter. The rim should be reinforced with a tough fabric (canvas) sewn in place to reduce wear and tear. The fabric should be wide enough (2 inches) to surround both sides of the bag border and space should be left inside the folded top for insertion of the frame wire. The wire should be about a yard in length. It is bent to accord with the circular bag opening as it is threaded into position. It is of sufficient length so that there will be around 3 inches of exposed wire at each end to be stapled to the handle. The handle end should be flush with the bag opening. To accommodate its diameter, make a V-shape cut to the fabric and stitch the edges. Use electrician tape to cover stapled areas.

Use of the Net. Sweep the net over the quarry with sufficient speed so that the insect ends up at or near the tip. At almost the same time, assuming it is near the tip of the net, quickly rotate the handle or flip the net so that the rim closes off the tip. Then place the opening against a smooth area of ground to prevent escape as you lift the tip of the net to allow the insect to crawl or fly upward toward the tip. When the insect is at the tip of the net, lift the net from the ground and slide a lidless glass gallon jar inside the net and move it toward the tip until the insect enters the jar. Temporarily close the jar by stretching the side of the net across the opening. With jar opening closed, insert the lid (it should have perforated openings for air flow) and slip it between the net and jar opening. Close the lid and remove the jar.

A crumpled piece of paper towel in the bottom of the jar will give the captive something to alight or crawl on and thus will provide it some comfort during it's brief captivity.

Photographing The Captive

As with the procedure with lizards, mark the capture site. Then transport the insect to an ice chest. After a period of cooling and signs of immobility, remove the jar lid to speed further cooling. In the meantime, select a place for the photograph. Ready the camera on a mock subject. If the insect is small, you will have to be ready for rapid-fire pictures because it will recover rapidly. Be careful with moths and butterflies to avoid damage to their thin delicate wing surfaces.

Saving Butterflies

Butterflies are declining in many areas of the planet, including the United States. There are both local and widespread efforts now to stem the tide. Not only are butterflies beautiful, they are also important pollinators. The decline of pollinators, which also include honey bees and some other insects, is an alarming trend because without pollination, many plants cannot reproduce—including many upon which we, and other animals, depend on for food.

In an article published by the National Wildlife Federation (Millar, 2008), titled "Restoring Rare Beauties'" ways are discussed that may help butterfly survival. Part of the effort involves planting food plants for butterflies, many of which are food specialists. Find out if there are species local to your area that need help. Maybe there is one in your yard? The caterpillar of the beautiful Monarch Butterfly, depends on milkweed plants *(Ascelepias)* for food.

See The Incredible Flight of the Monarch Butterfly. (KQED)

EVOLUTION: NATURE'S DRIVING FORCE FOR CHANGE

How organisms adapt to their environments

CHAPTER SIX

EVOLUTION: NATURE'S DRIVING FORCE FOR CHANGE

To be ecologically literate, and to combat "nature deficit disorders" (See p. 2) the process of "natural selection" should be understood. It is essential to an understanding of evolution. It explains issues of great importance to our future—among them why we are experiencing a population explosion and the power of exponential growth; how pest resistance to pesticides and increasingly more threatening human and other biotic pathogens emerge; why there are risks inherent in the release of genetically modified organisms (GMO's) and nanotechnology particles—particles in the size range of atoms and molecules. The rush to release them calls for more public oversight (See Gendler 2006, and Shand and Wetter, 2006 and 2006–2007).

Evolution is not a peripheral subject but the central organizing principle of all biological science. The recent history of the life sciences has increasingly demonstrated that "nothing in biology makes sense except in the light of evolution" (Dobzhansky, 1973).

Groundwork for understanding natural selection can begin in the upper elementary grades. Local populations of plants and animals can be studied to introduce the concepts of population, habitat, and environment. All students in middle school or high school should learn about natural selection. Topics to consider are individual variation including the difference between hereditary and non-hereditary variation, competition, exponential growth, the reproductive potential, environmental resistance and carrying capacity of the environment in relation to species population density. (See p. 179)

The Factors Involved in Natural Selection have been Summarized by Huxley (1966) as Follows:

1. All organisms show considerable variation.
2. Much of this variation is inherited.
3. In domesticated animals and plants, man is able to take advantage of inherited variation and produce new useful types by artificial selection.
4. In nature, all organisms produce more offspring than can survive.
5. Accordingly, there is a "struggle for existence"—not all the offspring will be able to survive or to reproduce.
6. Some variants have a better chance of surviving or reproducing than others.
7. The result of this is natural selection—the differential survival or reproduction of favoured variants: and this, given sufficient time, can gradually transform species, and can produce both detailed adaptation in single species, and the large-scale long-term improvement of types.

What follows are several activities and examples I have found effective in introducing evolutionary subject matter.

Learning about Exponential Growth (the Biokrene*)

Exponential growth (or increase) advances by doubling of numerical units: 2-4-8-16-32, etc. It can generate very large numbers quickly. Arithmetic, or linear, increase advances one numerical unit at a time: 1-2-3-4-5, etc., hence is much slower. Reproduction in plants and animals tends to be exponential, (to increase in geometric ratio). Exponential growth starts slowly but then accelerates and generates very large numbers rapidly unless held in check, in organisms, by a variety of environmental factors (environment resistance).

Thomas R. Malthus (socioecologist), in <u>An essay on the principle of population,</u> 1798, (1992 Reprint, Canbridge University Press) argued that since human populations can increase exponentially, they must inevitably outgrow their means of subsistence. The increase is sooner or later checked by hunger, disease, and war. However, he had no way of knowing that the advance in methods of agricultural production would temporarily greatly increase the rate of food production and delay his expectation. He became discredited. However, it is now clear that new "Malthus delaying" discoveries are failing to keep pace with the human biokrene, thus, in a sense, vindicating Malthus. Malthus' essay aided Charles Darwin in formulating his concept of natural selection.

---

*A term concocted by Professor Raymond Cowles to simplify references to the reproductive potential, the tremendous capacity for reproduction of virtually all living things (Bio) life and (krene) flowing well—"the flowing well of life." Cowles, 1977, Desert Journal, p.181, U.C. Press.

Activity: A Hands-on Reality Check

Bringing the sight, touch, and smell of fresh plant life into the classroom. Materials: scissors, white paper (8 1/2" x 11"), clipboard, and container for field-collected plant specimens. Make seed counts of individual plants and calculate reproductive potential of classroom samples.

I have worked with two widely available species that are especially suitable for this activity—the Wild Oat (*Avena fatua*) and the European Dandelion (*Taraxacum officinale*). There are surely many others. Check your local environment. Consider availability and difficulty (thorns, toxins, too many seeds, etc.) in making counts per plant.

Wild Oat *(Avena fatua)*

Its close relative, the Cultivated Oat (*Avena sativa*), is believed to have been derived from the wild species by early humans in the Middle East. It could also be used for this activity. The Wild Oat is a widely distributed weed in many temperate environments. Seeds are easily counted on fruiting stalks and totals per plant readily determined. Pull up plants by their roots, shake off dirt, and separate any individuals that may be growing close together. Give each student a plant for close examination and seed counting.

The number of seeds per pod (or spiklet) can be determined by counting the awns, usually two per spiklet. (see Fig. 43, p. 144)

With a hand lens examine the seeds and awns. Feel the edges of the awns. Awns have minutely saw-toothed (serrate) edges that are important in seed dispersal (see with hand lens). Wild Oat seeds and those of many other grasses attach to animals. Who has not had "fox-tails" in their socks or removed them from a dog or cat?

Determine the total number of seeds in the class population and the average per plant. Using the average, times number of individuals, calculate the reproductive potential of 5 generations of the class population reproducing exponentially at the average rate per plant.

The Wild Oat is a member of the worldwide Grass Family Poaceae with some 10,000 species.

European or Common Dandelion
(Taraxacum officinale)

Native in Europe and, like Wild Oat, also a widely distributed exotic. It is often found in lawns where considered by many people as a pest. Seeds per fruiting stalk are far more numerous than in the Wild Oat. My seed counts have ranged from 131–323 (7 samples). Individual plants may have one or two to many fruiting stalks in varying stages of development. With this species select one or two maturing fruiting stalks to simplify counting procedures.

Select stalks in the mature bud stage with the white tuft of the immature seed parachutes showing at the bud's tip. It is best to use buds because when the fruiting stalk has mature parachutes, the seeds are readily airborne. With scissors cut the bud in two about half way from the tip to remove the parachutes. Do not use yellow-tipped buds. The seeds may not be fully developed early in the plant's floral stage. (Fig. 45)

In making counts, open the cut bud by breaking apart the pedicel to which the seeds are attached. Remove the seeds and spread them out at the top of a sheet of white paper (8 1/2" x 11") attached to a clipboard. Be sure you have removed them all. Now test your power of estimation. Write down how many seeds you think there are. Then move the seeds downward, a small countable cluster at a time, and record each count in a vertical column at one side. You may need to recount a group several times to ensure accuracy. Accuracy is paramount both in science and everyday life. Counted seeds should be moved to the bottom of the counting surface. Do not include parachute stalks that lack definite seeds that have failed to develop properly. Finally, sum the counts to obtain the total. How close was your estimate? (Fig. 46)

Determine the classroom population's reproductive potential of the bud samples through several generations. See section on the Wild Oat (p.156).

Fig. 45. Left. Mature bud of Common Dandelion. A white tip indicates it is suitable for seed counting. A yellow tip indicates seeds may not be mature. Right. Bud in parachute stage.

Fig. 46. Common Dandelion seeds. Note group of seeds spread out for counting—recorded on the right.

There are many named species of Dandelion (*Taraxacum*), mostly reproducing clonally by asexual seeds. They are members of the large sunflower Family Asteraceae. Even if you have a mixture of species, they will still serve well to illustrate the reproductive potential of a highly productive, closely related group of plants (See Hickman, 1993 p. 350).

Some people may say why go to all the foregoing trouble of gathering plants and counting seeds in understanding exponential growth? Why not simply have students quickly calculate on their computers what happens with exponential growth?

EVOLUTION: NATURE'S DRIVING FORCE FOR CHANGE

When I was an undergraduate at the University of California, Los Angeles, I took, and later assisted, in a course given by Professor Cowles. The course was designed to introduce students to the flora and fauna of the nearby Santa Monica Mountains. Now, 60 years later, I vividly remember this classroom with its delicious sight and smell of the Santa Monica's—the California Sagebrush, Black Sage, Chamise, and other plants that brought wild nature to our senses.

Exponential growth can be quickly demonstrated with a computer, but this would contribute little to nature-bonding. Working with actual fresh plant life is more likely to do so. Through the reality of sight, touch, and smell, and hands on manipulation, brain cells are activated in ways no computer can duplicate.

Natural Selection: A Graphic Demonstration

This simulation is based on asexual reproduction to avoid the complexities of sexual reproduction. Participants should understand that sexual reproduction (in contrast to asexual reproduction) increases the genetic diversity of offspring and thus may often increase the chances, especially in changing and complex environments, of successful adaptive change (See Stebbins and Allen, 1975).

The demonstration was first developed, with the aid of my student Brock Allen, in the early 1960s when I was a member of the National Science Foundation's Elementary School Science Project of the University of California, Berkeley. Although more recent approaches have been developed, we have found that our exercise remains a useful teaching tool. When it first appeared, it was shown on TV and for over four decades was available in kit form, titled *Evolutionsspiel (The Evolution Game),* manufactured by Schlüter, Biologie in Europe·Haus für biologie, German Biological Supply House, Gerberstrasse·11 D-71364 Winnenden, Germany. The demonstration, as originally designed was used successfully from the 5th grade to, and including, the university level.

Activity: Selection for Concealing Coloration

A varied starting population of 100 "animals" (punched-out colored paper chips), 10 each of 10 different colors, are mixed and scattered randomly (using a shaker vial) on a multicolored fabric (about 4' x 3', the habitat) spread out on a table. The population at the start can be considered as near "carrying capacity" for the environment. Predation, one of the factors of "environmental resistance," checks population growth. Predators (birds, in the form of participants) on signal, catch prey (one chip at a time) until about 3/4 of the prey have been caught. Each predator might capture 18 or 20 chips (number to be designated by the supervisor of the activity (see below). A halt is called in foraging. The survivors are removed from the habitat and each reproduces (asexually), producing 3 of its kind, thus returning the population to approximately its former size. The population is then

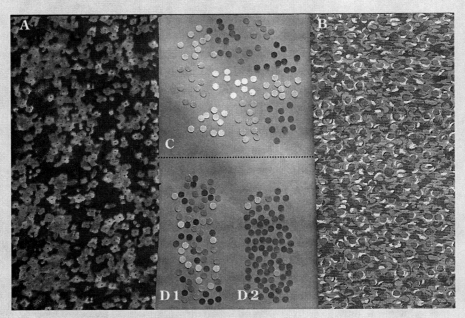

Fig. 47. Results of a natural selection demonstration showing the response to two different habitats, A and B, by the same starting population (top center C). Note below (left and right, D1 and D2) camouflaged populations after 2 generations.

returned to the habitat and scattered randomly as before. Foraging occurs again, and once more the <u>survivors</u> reproduce. After 2 or 3 such feeding episodes, the population almost always blends closely with its background. "<u>Natural selection</u>" for a match to background will have occurred. (See Fig. 47). It should be emphasized that the "birds" are to catch the prey at random and are not to search for particular kinds of chips.

Materials:
1. Book of many differing plain-colored sheets for punching out colored chips.
2. Standard paper punch that creates chips $1/4$ inch in diameter.
3. Several dozen transparent plastic vials with caps (around $2\,3/4" \times 1\,1/4"$)—containers for chips of different colors, sizes, and perhaps patterns.
4. Shaker vial (around $3\,1/4" \times 1\,1/4"$) with cap drilled at its center with a $3/8"$ diameter opening to be used in random scattering of chip populations onto the fabric habitat. Diameter of the drilled opening can be critical. Make it just slightly larger than the chips.
5. Fabric habitat(s) (around $36" \times 48"$) with colorful, complicated designs (fine-grained floral patterns work well). One for every four students.

Directions for conducting the demonstration:

Select a supervisor (student?) to oversee each demonstration. The supervisor starts and stops each exercise and guides participants through procedures.

1. Spread out the fabric habitat on a table and smooth out wrinkles.
2. Participants, usually no more than 4 to a habitat, 2 on each side, stand with their backs to the habitat while the supervisor, using the shaker vial, scatters the chips as randomly and evenly as possible over the habitat. A few moments may be taken to break up some excessively clumped chips.

3. The predators (birds?), still facing away from the habitat, are told to catch the first chip they see and, as they transfer the chip to their free hand, to quickly look away from the habitat, before capturing the next prey (they may do this as quickly as possible). This simulates the break in attention that often occurs as a predator pauses between captures. They should place captured prey in a container at each end of the table.

4. Depending upon the number of participants, the supervisor tells the players how many chips each should capture (around 18 to 20 with 4 participants) so that around 3/4 of the prey are caught. When all participants reach their quotas, the supervisor calls a halt to foraging and the survivors are shaken off the habitat onto the table. To avoid chips rolling onto the floor, lift the corners of the fabric, shake the survivors into the center and carefully invert the fabric as it rests on the table. Then, lifting the fabric by the corners again, and holding it close to the table top, shake it gently to be sure all chips have been dislodged.

5. Line up the survivors by color in a horizontal line spaced about 1" apart by color—for example 3 reds, 1 blue, 4 browns, etc. Each survivor now reproduces (asexually). Below each, place 3 chips of the same color in a column to restore the population to its former size of around 100.

6. All survivors and their offspring are now gathered, placed in the shaker vial, and the supervisor scatters them over the habitat as before. Again the predators are to face away from the foraging area. After the next round of predation, the population will usually have blended well with its environment. It has "adapted" to harmonize with background as a result of predation pressure. If it is desired to refine the color adaptation, repeat once or twice more the foraging-reproduction sequence.

7. Participants should sort chips and return them to their vials at the end of the activity.

EVOLUTION: NATURE'S DRIVING FORCE FOR CHANGE

Real World Examples of Natural Selection

Industrial Melanism

The process and outcome of natural selection, except in rapidly reproducing microorganisms, is seldom directly observed. Often the changes in a species are so gradual that they are not noticeable in the ordinary course of observation. However, there are exceptions. Such has been the case with the English Peppered Moth (*Biston betularia*). Populations of this moth live in non-industrial parts of England where they typically rest on tree trunks and branches that are coated with lichens, which give their surfaces a whitish appearance. The moths are predominately white, peppered with flecks of black, making them inconspicuous against the lichens. By virtue of a mutation, one or two black individuals appear, on average, in every hundred of these moths. Presumably because of their conspicuousness against the white lichens, however, populations of the black variety do not become established in these surroundings.

In industrial sections of England, however, the situation is different. There, industrial soot and fumes have changed the character of the forests. They have killed the lichens that once whitened the trunks of trees and darkened the bark with soot. The once-camouflaged white variety of the peppered moth is now easily seen against the dark tree trunks, and it is the black variety that is favored by the new surroundings. As a result the balance of population has shifted to the black variety and in industrial areas in various parts of England, the common peppered moth is now black, with perhaps only one or two white moths appearing for every hundred black individuals.

To test the hypothesis that the observed population shift was caused by the shift in background color, the naturalists N. Tinbergen and H. Kettlewell captured and reared moths of both varieties and deposited them in equal numbers on tree trunks of both sorts and released equal numbers of the moths in contaminated and uncontaminated woods. By direct observation of foraging birds, as well as by population counts, they confirmed the supposition. On a lichen-covered trunk, birds were observed to fly past white moths to pluck black neighbors. On a soot-blackened trunk the reverse was true. It was color rather than some other characteristic such as resistance to fumes that caused the ascendancy of the black variety in industrialized areas (See Kettlewell, H., 1958, Heredity, Vol. 12, Pt. 1 and 1959, Scientific American, Vol. 200, No.3). In a more recent study (Grant and Wiseman—2000) of the American Peppered Moth a decline in melanism occurred following clear air legislation! For additional information, visit http://www.millerandlevine.com/km/evol/moths/moths.html.

"Ring Species"—A Snapshot of Evolution in Progress

What are ring-species and what do they tell us about evolution? A ring species has a distribution shaped like a ring with terminal ends that overlap and act like separate species. Imagine a horseshoe with its free ends coming together. The ring may not be perfectly symmetrical and the populations that comprise it may be separated in varying degrees, but through genetic exchange, over time, they grade into each other as they diverge around the ring. The terminal forms, toward the ends of the horseshoe, may reach such a high level of genetic separation that they behave like separate species.

The example to follow, *Ensatina eschscholtzii*, a lungless salamander of the family Plethodontidae, sensitive to heat and dryness, is an example. It was "guided" in its distribution by the topography and climate of California. The low great Central Valley appears to have been inhospitable to salamanders for eons, but its bordering ring-shaped upland terrains have not. Note elongate ring-shape distribution of the salamanders surrounding the Central Valley, shown in white. (Fig. 48) Early in my zoological career I had the good fortune to work out much of the detail concerning the range of this animal and its geographic variation, based primarily on color differences (Stebbins, 1949). I also studied its natural history (Stebbins, 1954), helpful in understanding its evolution and dispersal capabilities.

EVOLUTION: NATURE'S DRIVING FORCE FOR CHANGE

Many evolutionary processes are illustrated in the Ensatina ring, thus it warrants discussion in some detail.

Hybrid Zones *(shown in black in the Sierra and Southern California)*

The plain-colored Yellow-eyed and the blotched Sierra Nevada subspecies coexist in the foothills of the Sierra where they hybridize extensively (note long black area). The Yellow-eyed form must have dispersed across the inhospitable Central Valley at a time of greater humidity than now (arrow on map Fig. 48 shows direction of the presumed dispersal route). None has so far been found there, although the town of Oakdale, along what I believe may have been the cross-valley route, seems a likely place for a remnant population as long as its island of oak woodland is not destroyed.

The plain-colored Monterey and large-blotched forms hybridize in Southern California in an isolated locality at Sawmill Canyon above Banning in the San Bernardino Mountains and in the uplands of the Peninsular Range to the south (note black areas on map). Here hybrids appear to be less common than to the north. One might expect greater genetic divergence with distance, but this needs more study. Indeed, at an area still farther south, at and near Camp Wolahi, so far no hybrids have been found. The Monterey and Large-blotched forms are behaving as distinct species. The gray areas on the map are intergrade or transistion zones, based on color characteristics, where one taxon (or subspecies) grades into another or shows features that appear to bridge the characteristics of the two taxa.

Incipient Species Level Breaks

Black dots in the *Ensatina* range indicate their location. They are breaks based on molecular studies that indicate areas that could perhaps justify formal species recognition for some segments of the ring. For example, there is an abrupt change in pigmentation in the vicinity of the Monterey embayment, between the Yellow-eyed and the Monterey form. The yellow eye patch suddenly disappears and orange belly color gives way to white in many individuals. Thus even on pigment data alone, well before the molecular explosion, I considered this possibility, but could see no value in doing so because it would obscure the evolutionary cohesiveness of this group of salamanders—so clearly united by color, anatomy, behavior, and ecological characteristics.

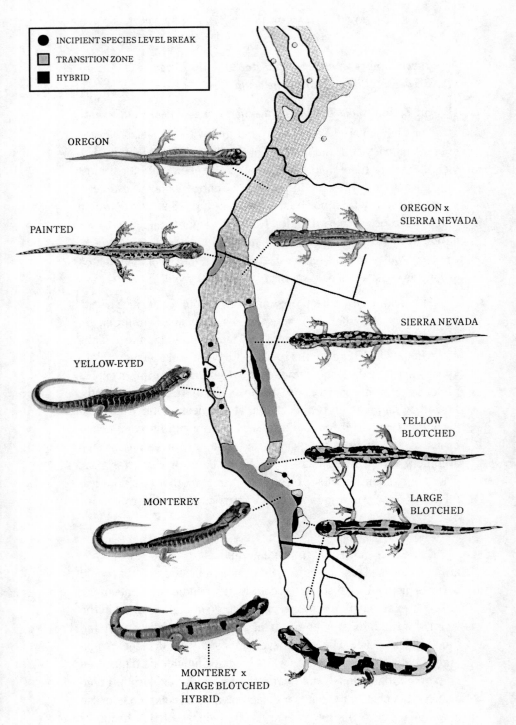

Fig. 48. The Ensatina Rassenkreis (Race Circle)

EVOLUTION: NATURE'S DRIVING FORCE FOR CHANGE

There is a gap in the range between the Yellow-blotched and the Large-blotch forms in the Transverse Mountain Range, a mountain system about 200 miles long. The gap extends from eastern Ventura County to western San Bernardino County, occupying a large part of the central portion of the range. Note arrow with attached black dot adjacent to the range of the Monterey *Ensatina*. Monterey *Ensatina*s are widespread, but spotty in the gap area. However, much effort by myself and others to obtain individuals of the blotched *Ensatina* have failed, despite the fact that there have been reliable reports at about the midpoint in the gap—one at Colbrook Camp in the San Gabriel Mountains and the other near, or at, Green Valley, about 30 miles airline to the northwest.

The evidence is strong that two interrelated factors may have severely impacted the blotched form in the Transverse Range: (1) competition from the upward spread of the presumably more dry-adapted, heat tolerant, unblotched form (the Monterey *Ensatina*) following increasing heat and dryness perhaps chiefly since the last glaciation, and (2) expected less tolerance of aridity and heat of the Blotched form, adapted to the lower temperatures and perhaps more persistent moisture conditions of its upland dispersal route. It now clings to pockets of habitat that appear to be less suited for, or to have escaped penetration by, the Monterey *Ensatina*.

Why then, with such a large gap in its range, should we not recognize the Large-blotch subspecies populations beyond the gap as a full species? I considered that option in my 1949 study and some other systematists have supported it, but I rejected the idea as noted above, because it would tend to obscure the cohesiveness of the ring's connections. Note the gray (remnant) intergrade sites (based on color) beyond the arrow that I felt tied together the Yellow-blotched with Large-blotched forms. The Transverse Range is characterized by extensively dissected topography and varied microclimates, rock formations, and vegetation. Pockets of Blotched *Ensatina*s may exist. Help is needed in searching for them.

Current studies in montane southern California concerning the distribution and taxonomic relationship of the large-blotched form *E.e klauberi* and the unblotched *E.e eschscholtzii*, calls for a delay in deciding their systematic status.

EVOLUTION: NATURE'S DRIVING FORCE FOR CHANGE

Perhaps an Answer for Those Who Question the Existence of Missing Links

Some people question evolution because each species may appear separatly created from all others. Where, they may ask, are the missing links? What follows may assist in clarifying this issue.

Fossilization is, with exceptions, a chancy event. Thus the record can be expected to have gaps. Nevertheless, there are, indeed, many known transitional forms. The *Ensatina* ring, however, is of special interest because we have "caught," so to speak, a species in the act of speciating. As things now stand, we have, intact, most of the links that hold the ring together. If a catastrophe were to occur that permanently wiped out all the members of the ring except the coexisting terminal forms—the Monterey and Large-blotched forms—probably virtually all systematists would agree that they would qualify as distinct species. That they are not so treated by some systematists is because of their connections around the ring.

We are fortunate to have this and a few other ring species with their overlapping terminal forms that contribute graphically to our understanding of the evolutionary process—in these cases, genetic divergence via distance rather than intrusion of more direct reproductive barriers.

PART TWO

NATURE BONDING

IMPEDIMENTS AND HOPEFUL PROSPECTS

THE HIGH COST OF ECOLOGICAL ILLITERACY

*The isolation of ecology.
Attitudes toward nature.
The population explosion.*

CHAPTER SEVEN

THE HIGH COST OF ECOLOGICAL ILLITERACY

In Part 1, I have focused on ideas and kinds of activities and observations that I hope may contribute to a harmonious relationship with nature. Now, in Part 2, I consider three major educational impediments that must be remedied to get us there. They are (1) isolation of ecology from mainstream education, (2) attitudes of domination or indifference toward nature and (3) overpopulation that results in increased strife, warfare, and environmental decline (this Chapter). The problems are interrelated and may seem insurmountable but they can and must be resolved. Historic and recent efforts show the way to a positive outcome (Chapter 8).

The Isolation of Ecology

As noted earlier, ecology can be a comprehensive science if so taught. In its broadest sense, it can embrace all the workings of nature from atom to cosmos and can shed light on all aspects of nature. Even space travel and the possibilities of extra-terrestrial life can come under its purview. It is an effort through science to understand how the intricate and interdependent relationships among living things and the physical world developed. Thus, <u>if designed to do so</u>, it can be *a much-needed unifying thread throughout public education*[1] (P. 2). (See also p.193)

Unfortunately, "Ecology has been isolated within biology departments as though it has nothing to do with the social sciences, the humanities, or other professions. The result is a *pervasive anthropocentrism* (italics mine) that magnifies the role of humans and their ideas, art, institutions, and technology relative to soil, water, climate, wildlife, resources, geography, energy, disease, and ecosystem stability" (Orr, 1992).

Hope for our future may, in a large part, depend upon the extent to which people, and especially their leaders, understand and act on principles of ecology and are ecologically literate (See Orr, 1992 and 2002 and Stone and Barlow, 2005). We owe it to our children to guide them in this endeavor.

Attitudes Toward Nature

An Excessively Human-Centered Perspective— Causes and Consequences

An <u>excessively</u> human-centered perspective can result in environmental problems: Some examples— (1) The assumption that humans have the right, or "manifest destiny," to extract from nature whatever is needed or desired regardless of, or indifferent to, the cost to other organisms. (2) Other organisms have little value except as they may in some way serve <u>our</u> needs and desires. "What good is it?" or "Of what use is it?" may be common questions. (3) Human ingenuity, technology, and know-how can ultimately solve all problems, including unending economic growth[2] (P. 2). <u>Nature exists to serve man, thus nature is to be used</u>. A giant sequoia is seen as board feet; deserts and swamps as places needing development. To what extent are natural resources considered as also belonging to the non-human animal and plant world?

"Not so long ago we were hearing about conquering Nature and the 'conquest' of space—as if Nature and the Cosmos were enemies to be vanquished (Sagan, 1997)."

The development of agriculture and the settled way of life began the separation of people from wild nature. Throughout most of human existence hunting and food-gathering societies, whose success depended on accommodating their lives to nature, usually saw themselves as part of nature and as existing at nature's sufferance. Nature was respected, feared, and worshipped, and is known to have been used by at least some groups on a sustained-yield basis, sometimes demanding strong measures to control population, dictated by the needs of survival[3] (P.2). Many native people expressed homage to the animals they killed. In those days when social groups were small, over-exploitation of food supplies and other resources sent direct signals to the tribe and, in time, contributed to built-in cultural mechanisms—often borne of dire necessity—taboos, rituals, and folklore, some of which acted as deterrents to population growth. Yet

THE HIGH COST OF ECOLOGICAL ILLITERACY

the collective demands of population pressure, just as today, must at times have overwhelmed their recognition of the need for ecological restraint (See Martin, 2005). Nevertheless, earlier and remnant groups of these societies had much to teach us about adapting to nature's requirements* (See also Anderson, 2005).

"Wilderness" is a term concocted by "civilized" humans. For aboriginal people it was home (Blackburn and Anderson, 1993).

"We did not think of the great open plains, the beautiful rolling hills and the winding streams with tangled growth as 'wild.' Only to the white man was nature a 'wilderness' and—the land 'infested' with wild animals and 'savage' people" (Chief Luther Standing Bear, Ogalala Sioux, in Nash, 1982).

An excessively human-centered focus can interfere with efforts at nature bonding. An African Bushman youngster would know where nearly everything needed in her or his world comes from and the effects of wanton destruction of animal and plant life. But many of us now seldom obtain feedback from our insults to the environment in direct personal ways as did people who live close to the land.** Some important environmental events, such as the overall effects of the decline in wild plant and animal life, that will greatly affect our lives may not be given the attention it demands. Economic and political forces sometimes act together to gloss over or hide what is actually happening. Lack of awareness of ecological processes and limitations combined with difficulties in separating questionable

---

* For example, both American Indians and Australian Aborigines mastered the use of low flame height fire in maintaining biodiversity and protecting forests through spot removal of the brush understory. It is only in recent years that modern forest managers have come around to the aboriginal approach (See Levy, 2005). For many years the "modern" approach had been to extinguish forest fires as quickly as possible, resulting in a build up of understory brush.

New Guinea Highlanders were able to live sustainably for tens of thousands of years before the origins of agriculture, and then for another 7,000 years after the origin of agriculture, despite climate changes and human environmental impacts constantly creating altered conditions. (Diamond, 2005 p.285)

** Because of increasing worldwide human density and rapid communication, things now belatedly, may be starting to change. Consider the great tragedy of the Katrina hurricane and other threatening climate events that appear to relate to human-caused global warming.

THE HIGH COST OF ECOLOGICAL ILLITERACY

science from that which has undergone rigorous peer review,* can obscure recognition of the seriousness of emerging social and environmental problems.

A nature-centered world view does not mean an abandonment of technology and an unrealistic "back to nature" approach. It does require, however, that our technology be used in a framework of ecological understanding and that our manipulative, tool-making capabilities, and scientific know-how, be applied in ways to accommodate ourselves to nature's requirements.

"The outstanding discovery of the twentieth century is not television, or radio, but rather the complexity of the land organism. Only those who know the most about it can appreciate how little is known about it. If the biota, in the course of eons, has built something we like but do not understand, who but a fool would discard seemingly useless parts? To keep every wheel is the first rule of intelligent tinkering."—Aldo Leopold, 1949.

---

* Peer review is an important part of the scientific process. It helps to uncover mistakes, biases, ambiguities, exaggerations, even deceptions in scientific pronouncements and publications.

THE HIGH COST OF ECOLOGICAL ILLITERACY

The Population Explosion

In the late 1930s, I was a young zoology student at UCLA, an assistant to Raymond B. Cowles, Professor of Zoology. One day we got on the subject of family. I remarked that I was the oldest of seven children and how happy my life had been with wonderful parents and my many siblings—that, therefore, I had resolved to have many children when I got married. He did not respond. I felt I had said something wrong. I had expected support, but he was unreactive and shifted the conversation to something else. I learned later that he was deeply concerned about overpopulation and its threat to everything humanity cherished. He did not wish to hurt my feelings with a negative remark.

Raymond Cowles was born of missionary parents at Adams Mission Station, Hluhluwe Valley, Natal, South Africa, December 1, 1896. His boyhood was spent in Africa when wildlife was abundant and he became enthralled with wild animals and fond of the native people (He became fluent in Zulu).

He came to the U.S. in his teens but returned to South Africa, repeatedly, over many years, even as he reached advanced age. Thus he obtained well-spaced attention-getting "snapshots" of environmental change over many decades in areas he had known since childhood, and witnessed, first hand, the growing impacts of increasing numbers of people on the land—deforestation followed by accelerating erosion, decline of wildlife, deterioration of native cultures and their livelihoods, and increasing strife. As a result, he became a champion and spokesperson for worldwide population control, before most other educators and scientists. His book, "From the Bondage of Human Numbers" never found a publisher. Why? To sample his deeply felt concern, see The Promethean Myth[4] (P.2).

I've wondered to what extent my own great concern over excessive population growth is primarily the result of his influence or whether I would have arrived there anyway as I studied biology. I suspect the latter, but he brought the matter to my attention early, and into sharp focus.

THE HIGH COST OF ECOLOGICAL ILLITERACY

We have effective and humane methods of birth control yet why are so many modern societies failing to act effectively?

Statistics present a sobering picture but there is hope that we may see a turnaround. It almost happened in the 1950s to 60s (See p.183). Barring a great increase in mortality and/or birth control, population has been projected to reach 10-12 billion by the end of the twenty-first century. However, a recent estimate by the United Nations posits 8.9 billion by 2050, down from around 9.3 billion projected previously, reflecting a slowing growth rate. Some of this decline relates to the AIDS epidemic*. Population is currently growing at a rate of around 77 million people yearly (2001 World Watch estimate)—every three days increasing by about a half million people.

Humans and the Biokrene

During our long evolutionary past our species has surely undergone many fluctuations in population size as a result of climatic changes, variation in food supply, competition, strife, entry into new areas, and other factors, in the same way as other animals. As with other organisms, population growth leading to the filling of all usable environment (the force of the biokrene, p. 155) has, for the most part, been unrelenting. Thus, such population pressure among humans is not simply a phenomenon of the present age, but has frequently been with us but now is increasing nearly everywhere on earth and there are few new lands left to conquer. The population pressure engendered by high population density is a major contributor to strife, violence, warfare**, spread of disease, and many other complex social

---

* "The U.N. projections, while promising, are not guaranteed. More than 150 million couples worldwide still want to limit or space their pregnancies but lack basic contraceptive services. This is not just a problem in other countries, such as India or China, but in the United States, too. The U.S. is one of eight countries expected to contribute to half of the population increase." In addition to the U.S., India, and China, there is also Pakistan, Nigeria, Bangladesh, Ethiopia, and Democratic Republic of Congo. (From the United Nations Population Division, World Population Prospects: The 2002 Revision)—Gloria Feldt, President of Planned Parenthood Federation of America, in Planned Parenthood Tomorrow, No. 13, Feb. 2004.

** The Rwanda Hutu/Tutsi fighting and genocide in Africa, declared as ethnic strife, appears now to have resulted mainly from population pressure. (See Remember Rwanda? by James Gasana, World-Watch, 2002, Vol. 15, No. 5)

THE HIGH COST OF ECOLOGICAL ILLITERACY

and environmental problems. It is a serious impediment, if not an outright barrier, in many countries now to economic development[5] (P.2).

It is important in considering problems of population growth to recognize Liebig's ecological "Law of the Minimum"— that of the multitude of factors <u>required</u> for an organism's survival only <u>one</u> need be in deficient supply to result in its demise. The great "unused" spaces in Nevada and other western states, often pointed to as evidence that we, in the United States, are far from being overpopulated, give an ecologist little comfort. Their carrying capacity for humans is low.

<u>Carrying capacity</u> needs to be the focus in considering overpopulation—not population density alone — namely, how many individuals can survive <u>over the long term in a given area</u> without degrading resources needed for survival, and <u>without input of resources from outside areas</u>. We must also allow sufficient leeway between our numbers and our environmental requirements to allow for fluctuations in environmental conditions.

The desire of women throughout the world to control their fertility and reproductive health must not be underestimated. It extends beyond ethnic, economic, educational, or religious status (See Snell, 2004). Where reproductive health care services are not available or restricted, self-induced abortion and spread of reproductive diseases, with their often dire consequences, attest to their desire to control their reproduction. It indicates that reducing poverty and improving general education, as important as these actions are, they are <u>by no means</u> the only route to birth control. Specific guidance on reproduction, and access to birth control methods and devices are also required.

In the absence of adequate reproductive health care, there are further problems. Preventable premature deaths of parents (from HIV and other diseases) results in children denied adequate nurturing and care, and to youthful alienation, crime, and violence. It is estimated that three billion people—roughly half the Earth's total population are under the age of 25, with all or most of their reproductive years ahead of them and without much information or guidance on sexuality or reproduction (World Watch, May 2004). An added concern is bias toward male offspring, which results in a preponderance of males with their unresolved employment and testosterone needs that can contribute to strife.

Unfortunately, the pressing international need for worldwide, comprehensive, well-supplied and abundant reproductive health care clinics currently remains tragically under-supported and under-funded (See Snell, 2003). The problem also exists in the United States! It is now critical that this problem be dealt with as <u>rapidly as possible</u>. There is no more important issue relating to the survival of human and other life on this planet! It is a failure that has often happened in the past, but now we have no other place to go (See Diamond, 2005). If the idea of world population control is not realistic, then there is only one realistic view of the future: nature will, indeed, have the last word.

The United Nations Environment Programme (UNEP) claims "the natural environment will be increasingly stressed if present trends in population growth, economic growth and consumption patterns continue. The continued poverty of the majority of the planet's inhabitants and excessive consumption by the minority are the two major causes of environmental degradation. The present course is unsustainable and postponing action is no longer an option." (See World Watch, 2004, Vol.17, No.5).

HOPEFUL PROSPECTS

An historical message that almost succeeded.

With special thanks to Margaret Sanger.

CHAPTER EIGHT

HOPEFUL PROSPECTS

In the 1950s and 60s, largely through the persistence, persuasiveness, dedication, and funding skills of businessman, Hugh Moore, birth control was further removed from the shadows and became recognized as absolutely crucial to human survival. He was able to enlist scientists, business leaders, industrial and financial experts, top politicians, including presidents Nixon and Johnson, in support of the cause. <u>Rampant population growth became ranked with nuclear disaster as the greatest threat to peace on earth</u>.

A Hugh Moore Fund was created and its primary target was to get the Federal Government to commit to population control. Nothing less than that was required to achieve success. Moore emphasized that foreign aid to help the needy and starving in the world was being nullified by the population explosion. John F. Kennedy was the first U.S. President to concern himself officially with the problem of population limitation. But it was President Nixon who took the first courageous step in addressing the population issue. He persuaded Congress to create the Commission on Population and the American Future (the "Rockefeller Commission"), which concluded it could see no advantage in further growth of the American population. Unfortunately, he shelved the report for political reasons (as have evidently all subsequent presidents). However, as a result of his leadership, the U.S. Senate conducted extensive hearings on the subject (See Bibliography—Population Crisis 1965-1968). Lyndon Johnson declared "Let us act on the fact that less than five dollars invested in population control is worth a hundred dollars in economic growth." Foreign aid was going down a rat hole. A Gallop Poll asking whether birth control should be made easily available to any married person wanting it was supported by 81% of Catholics and 86% of non-Catholics! But there was inadequate support from the Catholic bishops and an appeal to Pope Paul VI by 81 Nobel Laureates, 36 from the U.S., to modify his position on birth control also failed (See Bibliography Population Crisis Part I, 1965-1968).

To the concern of businessmen who worried that controlling births would hurt business, Moore responded that the poverty caused by population growth would eventually create a public incapable of buying.

How different is the outlook today, over three decades later! The problem has grown far worse in the interim and, sadly, it has fallen off the government and to a large extent the public, radar scopes in the United States.

The unresolved urgency for meaningful action on this matter was highlighted by the 2002 United Nations World Summit on Sustainable Development at Johannesburg to which, regrettably, the United States failed to offer effective support. The official delegates concluded their 10 days of deliberation on September 4th without addressing, in any significant way, the problem of overpopulation, which if not solved, all other efforts are bound to fail (See Mann, 2002).

This is deeply saddening in view of the above great effort to galvanize action in the United States and abroad to deal with this serious global threat to humanities' future (See Lader, 1971, Breeding Ourselves to Death). Had the U.S. senate hearings and recommendations that followed in the late 1960's been implemented, they would have lead to positive change. However, at least our government thoroughly faced the issue then. In the meantime, world population has doubled, having gone from around 3 to 6 billion people and accompanying environmental and social costs have greatly escalated.

Despite the fact that polls show the American public is not enamored of further population growth, there is a virtual political taboo, because the progrowth argument is endorsed by the powerful (Grant, 2004). We are at a tipping point now on the population issue and its moral implications concerning birth control.

HOPEFUL PROSPECTS

Some Reasons for Hope

Humanitarian Efforts of the Gates Foundation

Established by Bill and Melinda Gates of Microsoft. The foundation is dedicated to improving Third-world health, agriculture, living conditions, and education—in a major effect to lift people mired in poverty and disease (including the HIV/AIDS epidemic) out of their tragic conditions. The Gates believe that saving peoples' lives through improving their food supply, education, and health translates to population stabilization. The foundation is prepared to use much of its wealth to achieve its humanitarian goals.

Enter a second player in this remarkable dedication—businessman Warren Buffet who believes in returning much of his extensive wealth to the benefit society. The Gates and Buffet have created a merger that has the potential of producing a great and lasting gift to humanity. We may hope for a significant increase in support for medically comprehensive public health clinics throughout the areas they are targeting. Mrs. Buffet (deceased) was very concerned with family planning and reproductive rights.

Iran's About Face on Family Planning

Under Ayatollah Khomeini's leadership in 1979, family planning programs, put in place by the Shah in 1967, were dismantled. The Ayatollah wanted large families to provide soldiers for the war with Iraq (1980-88). Fertility levels increased to near biological maximum with the result the country's leaders realized that overpopulation was resulting in serious environmental degradation and unemployment, undermining Iran's future.

Therefore, in 1989, family planning was restored and in May 1993 a national family law was passed. There was mobilization of agencies to encourage smaller families. Some 15,000 "health houses" (clinics) were established to provide rural populations with health and family planning services. Religious leaders were involved in a "crusade" for smaller families. A full array of contraceptive measures, including male sterilization, was introduced. All forms of birth control were free of charge. Indeed, Iran became a pioneer—the only country to require couples to take a class on modern contraception before receiving a marriage license.

A broad-based effort was launched to raise female literacy. Between 1970-2000, female literacy increased from 25 to 70 percent. TV was used to disseminate information on family planning throughout the country. The result was (from 1987-1994) Iran cut its population growth rate in half. Its overall rate of 1.2 percent in 2004 is only slightly higher than that of the U.S.

If a country with a strong tradition of Islamic fundamentalism can move quickly to population stability so can many other countries. Countries everywhere have little choice but to strive now for an average of two children per couple. There is no feasible alternative if we are to have a safe future (See Plan B 2.0 by L. Brown, 2006, Chapter 7 Eradicating Poverty, Stabilizing Population). When women are better off, so are populations.

Jimmy Carter's book (2005) "Our Endangered Values—America's Moral Crisis"

In it he supports abstinence prior to marriage but, <u>at the same time</u>, recognizes the importance of teaching medically approved information on birth control and reproductive health threats. See also World Watch, September/October 2008 (Bilbliography, p. 253). When women are better off, so are populations.

It is now absolutely essential that the insightful and remarkable effort of the past, energized by Hugh Moore, must not be lost. We should learn from it and revitalize its life-saving message.

"There is no human circumstance more tragic than the persisting existence of a harmful condition for which a remedy is readily available. Family planning, to relate population to world resources, is possible, practical and necessary. Unlike plagues of the dark ages or contemporary diseases we do not yet understand, the modern plague of overpopulation is solvable by means we have discovered and with resources we possess. What is

lacking is not sufficient knowledge of the solution but universal consciousness of the gravity of the problem and education of the billions who are its victims."

From a speech written by Martin Luther King Jr., delivered May 6, 1968 by his wife Coretta Scott King.

A Land Ethic

We must strive for a land ethic to ensure sustainability. Its ecological principles are well established and based on solid science.

Ecologically, according to Aldo Leopold, the Land Ethic entails a limitation on freedom of action in the human struggle for existence. It recognizes that the vast majority of other organisms, whether we regard them as beneficial or not, have a right to exist and that we have a moral obligation to protect them, at some self-sacrifice if necessary. The ethic should be embraced whether or not one cares about other life—as a matter of survival. In view of our present reproductive success and ability to modify the earth's environments, continued unlimited pursuit of immediate human interests and needs at the expense of other organisms will ultimately result in the overpowering of natural ecosystems and widespread major disruption of the biosphere. Our own existence will then be in jeopardy.

In short, a land ethic changes the role of *Homo sapiens* from conqueror of the land-community to plain member and citizen of it. It implies respect for his fellow-members, and also respect for the community as such (Leopold, 1949).

PART THREE

EDUCATIONAL PRIORITIES

ECOLOGY AND THE
ENVIRONMENT—
A UNIFYING THEME.
A NATURE-CENTERED
EDUCATIONAL PROGRAM.

EDUCATIONAL RESPONSES

Ecology and the Environment—A Unifying Theme in Education. Integrating Departments of Science and Education.

CHAPTER NINE

EDUCATIONAL RESPONSES

Since the key to success of the nature-centered ecological program depends so heavily on direct personal experiences with living organisms and their habitats and on doing studies outdoors, it is important that ample time be allowed for the program, especially at the primary and middle school levels when children are at a formative stage. This can be fully justified because later on, in addition to the scientific and attitudinal goals, the program can make important contributions to virtually all other disciplines including writing and art. It can be made highly interdisciplinary.

"If you take ecological principles as your guide, you can't study things in parts—there's a disconnect, an imbalance of information that keeps you from understanding the whole"
—Conrad Benedicto, high school teacher (See Snell, 2003, p.39).

However, some existing ecology courses may need to be adjusted to meet the goals expressed or similar alternative course work designed to do the job (See discussion of our MVZ natural history course, p.195–196).

Thus ecology, so designed, can provide a framework for relating many different fields of knowledge—biology, physics, chemistry, history, geography, civics, economics, and others. These subjects need not be taught as if isolated. All address some aspect of the physical or biological characteristics of the earth and can relate to past and present human actions and their effects. Thus they all can connect with ecology and the environment, and can participate in the goal of achieving a nature-centered world view. A special effort should be made to relate subject areas to this important goal. For example, what have been some of the effects of wild organisms and natural environments on the course of history? Economics should embrace ecological principles and methods to achieve sustainability1 (P. 3). Chemistry and physics can be enriched with examples from biology--the form and function of sense organs and other anatomical structures that have evolved in response to the chemical and physical properties of nature.* Geology and geography can draw extensively

---

* This could be accomplished by an occasional interchange of lecturers that have met to integrate their presentations.

EDUCATIONAL RESPONSES

on the role of organisms in shaping the constituents of soil, the atmosphere, etc. and human distributions and cultures.* The teaching of languages can help by employing nature topics in the development of writing and speaking skills.

A current trend in education places emphasis on the development of thinking skills—creative thinking, critical thinking, and problem solving. Training in the scientific method and the observational techniques of a nature-centered program tie in with this goal, as well.

Establishing and Maintaining the Nature-centered Educational Program

The Role of Higher Education

Given the serious social and environmental issues now facing us, biological studies <u>that include ample time for field work</u>, should be <u>required</u>, at some point, in the training of primary, middle school, and secondary school <u>teachers</u>. For teachers without such training and those desiring "refresher" studies, appropriate workshops and course work should be made easily available and, some salary incentive offered. Course work in biology for both majors and non-majors should include a broad ecological/evolutionary and "humans as a part of nature" orientation, with important principles of biology presented in evolutionary context. <u>It should involve the student in doing science</u>, through personal participation in research, rather than just learning about science. Practically all science study deals with known facts and neglects the processes by which these facts are obtained (Leon Hunter).

---

* There are many TV history programs now that are connecting historical events with environmental and social changes. A recent one focused on the Little Ice Age and the relationship of nature's vicissitudes to famine, warfare, changes in agriculture, and human cultural activities.

Thus teachers should have some experience with doing field research in order to have first hand experience with the scientific method as applied in outdoor studies (See p.198, Roving Professional Naturalists).

The course should cover the spectrum of biological organization from molecules to ecosystems and the biosphere. However, it should not overdo molecular aspects at the expense of the ecological because the latter is so essential to establishing commitment to the protection of nature and our social well-being. Regrettably, there still lingers some intellectual snobbery among biologists at both ends of the biological spectrum. I trust it will soon completely disappear. "Biology is indivisible; biologists should be undivided" (Bartholomew, 1986; and see also Simpson, 1969).

Comments on Teacher Training

If a teacher does not have enthusiasm for the study of nature outdoors, that will be conveyed to the students, so it is critical to the nature-bonding program that training be designed to impart such commitment.

For the general high school biology course, I recommend the Biological Sciences Curriculum Study (BSCS) "Green Version"—an ecological approach, as an ideal text, supportive of the above goals. It is widely used in the U.S. and abroad and has often been the choice in "3rd world" countries (See Bibliography).

Training for teachers in outdoor ecological/environmental studies is more difficult to prescribe. Something like our course in Vertebrate Natural History at U.C. Berkeley would be ideal for teachers and we have had some in our classes but it was chiefly designed for undergraduates and some graduates in zoology. Joseph Grinnell, founder and first director of U.C. Berkeley's Museum of Vertebrate Zoology, originally created the course and it is still being taught in our Department of Integrative Biology.

EDUCATIONAL RESPONSES

Regrettably, and to the disservice of the general public and the goals I espouse, natural history courses have been considered "old fashioned" by many science departments and have languished or been abandoned. Now they should be reinstated and reenergized because all teachers need the field orientation they contain.

I've tried to come up with another word other than "history" but so far have failed. It's a misnomer. The subject matter of natural history goes <u>far beyond</u> history. It is akin to ecology. However, natural history studies, perhaps, tend to be more preoccupied than traditional ecological investigations—with species identification, systematics and distribution, observational techniques, opportunistic observations of species behavior and interactions in nature and, at least as taught at U.C. Berkeley for many decades, the impact of humans on the biosphere.

Natural History at Berkeley requires the students to spend 4 hours in the field each week in group sections under the supervision of a faculty member or teaching assistant. Focus is on the "terrestrial" vertebrates—amphibians, reptiles, birds, and mammals. Species recognition, their habits, interactions, and their place (or role) in nature are emphasized. Repeated visits are made to several different habitats in the East Bay Hills, so the students come to know many of the physical and biotic features of these areas and develop a sense of place. They are trained in techniques of observation and careful, on the spot, note-taking. A 3-hour lab each week involves study of preserved specimens and sometimes live animals, including demonstrations of behavior. Lectures are given twice a week. At the end of the course, each student is required to turn in a personal report on a field research topic of choice and his or her field notebook for evaluation.

Throughout a period of over 30 years enrollment grew from around 25-30 to over 125 students, at which level we had to put a cap on enrollment.

One of the attractions was being able to have live animals in the laboratory, now greatly constrained by concerns over collecting and laboratory care, but the free-living wild animals are still the biggest attraction! As one of our students who started out with little interest in birds said to me at the end of the course, "Dr. Stebbins, this course has finally hooked me, I can't go anywhere now without noticing the birds." Direct experience with wildlife, however small or large or where ever found, is a key element in nature-bonding. "A bird in the heart is worth more than 100 in a notebook." (A quote from my father)

Many other courses surely qualify in providing outdoor experience with focus on the natural environment—insect biology, marine biology, plant taxonomy, botany, paleontology, anthropology, ecology and environmental studies, geology, geography, etc.

Credit for outdoor environmental studies could be given for volunteer work with wildlife agencies—the National Parks, Municipal Watersheds, Natural History Museums, conservation organizations, etc. Of primary importance is that the teachers in training have on-site experience in the field with some professional guidance, in settings that will foster nature-bonding (See below).

Departments of education, the sciences, and humanities should cooperate to develop, through experimentation and course work as necessary, subject matter and approaches in teaching that will assist the nature-centered educational program. On the U.C. Berkeley Campus the Lawrence Hall of Science, its adjacent Lawrence Hall Nature Area, and the Botanical Garden, adjacent to wild lands on the campus, in location, expertise, and experience, have offered such an environment.

In the 1960s Professor Lawrence Lowery, of the Department of Education, and I collaborated in teaching such a course. Our students were drawn from the U.C. Departments of Education and the Life Sciences, and our experimental subjects were ethnically diverse urban elementary school children from Berkeley. Our study area was the Lawrence Hall Nature Area. Our students worked with the children, turned loose in the natural environment, to get ideas for projects that might support our goals of getting people connected to nature. Many innovative activities were devised[2] (P.3).

EDUCATIONAL RESPONSES

Faculty and students in education, biology, physics, chemistry and mathematics have, for many years, interacted with each other and visiting school children and their parents at the Lawrence Hall in furthering science education. Such cooperative educational mixes should become widespread and active participants in nature-centered programs.

Roving Professional Naturalists

Teachers and students who have never conducted a scientific study of <u>free-living animals or plants outdoors</u> will be greatly helped in learning the scientific method, as applied to such studies, by having their first attempts supervised by experienced persons. The participants should be encouraged to formulate questions and methods to seek answers, but with professional help as needed. Such assisted observational or experimental studies speed up their independence in future investigations (See p.82 under Barstow School District—problem solving for teachers). Elementary schools in Vernon, British Columbia, Canada, for many years had access to such a person, naturalist Jim Grant, who helped teachers with nature study programs. Such a person can also assist in setting up classroom activities (see p. 36–38). Regrettably, at the time of this writing, the position no longer exists.

If we are to close the widening gap between civilized humans and nature, we must find ways to support such professionals as an integral part of the nature-bonding, whole organism approach to biological education. Continual in-service training of the teaching staff is essential to getting students involved in environmental education (Leon Hunter, p.82–83).

The Importance of Political and Administrative Support

Success of the nature-centered program will depend on the degree of commitment to it within the political and educational systems themselves. Unless school administrators, school boards, and key state and federal authorities are convinced of its value and are behind it, chances of widespread success will be slim. Only highly

EDUCATIONAL RESPONSES

dedicated teachers are likely to act on their own. Teaching strictly to meet requirements of a syllabus or other closely prescribed directions, can kill innovative programs. They simply can't survive if administrators, teachers, and students want only to ensure high test scores that may or may not address the concerns expressed in this document. We must work for greater flexibility. Indeed, after the 3Rs, and professional training, what is most important for the majority of humanity? A strong case can be made for <u>ecological and environmental</u> literacy. For the program to work it must ultimately be incorporated within state and national educational standards.

There are those in high places in our government (the U.S., 2003) now who argue that environmental education is "ineffective" (See Snell, Sierra, Nov./Dec., 2003, p.38). Whatever the source and its motivation, it sends a mixed, regrettable, and probably inaccurate message that should not be allowed to deter us. If, indeed, such education is "ineffective," every effort should be made to ensure that it <u>is</u> effective because ecological literacy is so important to our future.

The nature-centered program must not be looked upon as "just one more" subject area trying to find a place in a crowded school curriculum. Its importance is far too great for this. Ways must be found to ensure that ecological, nature-bonding activities are being taught and that ample time is available for studies in depth. School or district science coordinators, roving naturalists or other experienced field research specialists with teaching skills should be made available to keep the program "high-profile", conduct in-service workshops, help with arranging field trips and "walk-through" research projects for teachers. They can also help with reference materials and the establishment and use of school outdoor study areas—garden plots or off-campus study sites. Investment by government and the public schools in the staffing and maintenance of such facilities will reap great and long-lasting benefits for society. (See 3 (P.3))

CONCLUSION

CHAPTER TEN

CONCLUSION

Now, more than ever before, we need a deepening concern for the well-being of others, not only those within the circle of our relatives and friends or the group to which we pledge allegiance, but to all humanity. Further, we must also greatly expand this concern to include more fully the non-human living fabric of this planet in all its diversity and beauty.

A great challenge is before us. Can the scientific communities, great religions of the world, military establishments, the business and commercial interests of consumer societies, and growth-oriented economic and political structures embrace the change in thinking and action now so urgently required (See Mathews, 1990)?

I cannot accept the premise that things have gone too far and that nothing can now save us. There is a developing groundswell of hope that must become the great force of the future. What we are short on is time. Given enough time, and if we can escape nuclear and eco-catastrophe, a more humanely oriented use of science, I trust, will in the end prevail.

With electronic media, messages travel fast. There are magnificent science and nature programs now appearing on public television, and many dedicated people and organizations working throughout the world toward the spread of ecological understanding, environmental protection, and the preservation and restoration of wild animal and plant life. Wildlife protection, for many people now, has become an international goal. Indeed, this is one of the common goals that can help defuse violence and bring about international cooperation. Look at what schools have been doing in the California Desert (See Fig. 49). The map shows the many places visited primarily for biological field studies in the course of a typical school year in the mid-1970's. An update survey would be desirable.

I have a vision that in the vastness of space there are many other worlds where life has evolved and where thinking, reasoning beings resembling ourselves exist. On chance alone, some of them must be occupied by intelligent beings that early in the development of their civilizations, retained their humility toward nature, and with it recognized the importance of controlling the size of their populations and tool-making capabilities, who found ways to reduce selfishness

CONCLUSION

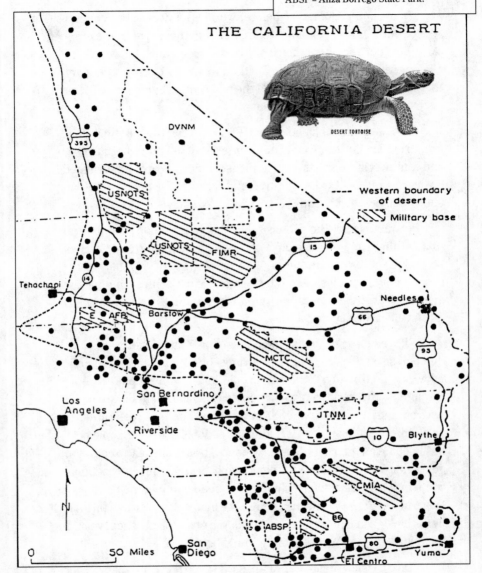

Fig. 49. Sites visited by educational institutions conducting teaching and/or research in the California Desert (1975–1977). A total of 272 sites were reported. In addition, some schools indicated general use of the desert. Because of the scale of the map, not all sites are shown. Included in the survey were a few semiarid localities outside the official desert boundary, along its western edge. A complete listing of all sites was sent to the BLM California Desert District office in Riverside, California. Map prepared by Gene Christman, Staff Illustrator, University of California Berkeley, Museum of Vertebrate Zoology. Both Death Valley and Joshua Tree are now National Parks.

CONCLUSION

and the violence of their fighting tendencies, and who learned to use their world on a basis of continual productivity—their cities and farms separated by large areas of little-disturbed natural countryside. They have technological and social amenities that resemble ours, but by exercising restraint have achieved them without jeopardizing the natural ecosystems of their planet. They are the advanced civilizations in the Universe. It shall be my dying wish that Earth may yet find a way to be among them.

"We travel together, passengers on a little space ship, dependent on its vulnerable reserves of air and soil, committed for our safety to its security and peace, preserved from annihilation only by the care, the work, and, I will say, the love we give our fragile craft. We cannot maintain it half fortunate, half miserable, half confident, half despairing, half slave to the ancient enemies of mankind, half free in a liberation of resources undreamed of until this day. No craft, no crew, can travel safely with such vast contradictions. On their resolution depends the survival of us all."

—Adlai Stevenson, 1965

CONCLUSION

To Mothers Everywhere

This song was written by our daughter Mary to her son Gabriel. She expresses her love for nature and her deep-felt desire to share this love with her child.

Gabriel's Song

CHORUS

Gabriel,
I brought you to the mountain
Before you were born
I bring you again in my arms.

Gabriel,
I give you the rock and the sky and the sun
I give you the mountain
My dear little one.

Your aspen leafed childhood I lay in the sun
New grass mirrored raindrops you are.
I stand to one side
With the ancient green pine
Your childhood a memory
Of blue-eyed time.

CHORUS

Your cradle a rock, glacier carved, water smoothed
A thousand years of the sun is your bed.
I stand to one side,
I watch as you sleep
Your summer green childhood
A memory to keep.

CONCLUSION

Dear Gabriel,

When you were a tiny baby, less than three months old, I took you camping at the Illecillewaet Glacier. I carried you in your snuggly nestled against my breast, up and over the glacier carved rocks, to a resting place where I laid you in the sunshine. Months earlier, while you were still in my womb, I had come to this same place on cross-country skis. When I thought about what it was that I could give you, I felt one treasure I could offer you was my love for the earth, the mountains, the rock, the sky, the sun, just as my father, your grandfather, had offered these gifts to me. There, beneath the mountain, I wrote you this song.

— *Mom*

APPENDIX
Other Successful Models for Teaching Nature Study

East Bay Municipal Utility District Biological Survey	210
East Bay Park's Naturalist Program	213
Orinda School District and Wagner Ranch	217
Mojave Max	220
Understanding Ecosystems from the Ground Up	225

APPENDIX

East Bay Municipal Utility District Biological Survey

By the Author with guidance from the then EBMUD Manager Steve Abbors

Municipal watershed lands that are open to the public can provide excellent opportunities for nature-oriented education for urban and suburban populations. Many are open for boating and fishing. They can also help people connect with wildlife and nature.

The EBMUD watersheds, located in the hills east of San Francisco Bay, California, together consist of 28,000 acres of public land rich in wildlife and surrounded by urban and suburban developments. The term "wildlife," as used here and elsewhere, includes both plants and animals.

In 1995 the board of supervisors decided that maintaining the biodiversity of its holdings would be the focus of management to ensure a high quality water supply for its constituents. Mr. Abbors, an outstanding professional naturalist, had much to do with setting this goal.

They embraced the viewpoint that *"complex ecosystems, with their diverse array of plant and animal species and their many biochemical pathways, provide a variety of buffers that impede or slow the rate at which pathogens and toxics, precipitated from the atmosphere or from other sources, can reach reservoir waters."*

Management was to be guided by this theme, and a process began to try to ensure this goal. The biota was to be studied in some detail to help achieve its long-term integrity. This required information on species and their community compositions, the role of the key wildlife players, their interactions over time, and the nature of human impacts and how best to manage these factors over the long term.

In 1996 two documents were published that resulted from the preliminary work: "Biological Survey Studies for the East Bay Municipal Utility District: Guidelines I. <u>Gathering and Recording</u>

APPENDIX

<u>Wildlife Information</u> and II. <u>Species Lists and Maps</u> (See Stebbins, 1996)." The management plans at this stage were experimental and subject to changes with experience. However, to put them to the test involved both staff and public training sessions (a half day on Saturdays). Their purpose was to inform participants of project goals and procedures in recording wildlife information to ensure standardization for computer processing.

Mapping of the location of individuals of species targeted for study was a key feature of the program. The locations for the sightings were entered on maps that were made available for all parts of reservoir holdings. Maps were based on aerial photographs that included contour lines, vegetation areas, roads and trails, and a variety of other landmarks. On the back of each was a series of columns for recording information for each sighting, animal or plant.

From the start it was evident that, for management purposes, it would be necessary to greatly limit the number of species to be given special attention. For example, at the time of publication (1996) the count of flowering plant species stood at 684 revealing a remarkable diversity! The list continues to expand.

For management, therefore, four species categories were established: (1) legally protected species, (2) keystone species—those with many dependent species, (3) feral or pest species, and (4) indicator species—those believed to indicate environmental trends, such as frogs. Long term field studies are particularly critical in assessing ecological trends and problems (See Note 2, Part 1). However, for purposes of public education and the development of nature bonding, people were encouraged to also target whatever they wished.

There has also been focus on grazing management using livestock to simulate, so far as possible, the foraging patterns of native grazers, many of which have disappeared. Cattle are rotated on the range so that impact on native grasses and forbs (wildflowers, etc.) is minimized, permitting plant diversity to expand and flourish. By timing cropping before seeds set, the dominant non-native grass, the Wild Oat (*Avena fatua*), that crowds out grasses and other native plants can, hopefully, be expected to decline.

APPENDIX

Grazing level is also considered in relation to its impact on the widespread native grazer, the California Vole (*Microtis californicus*), an important constituent in the food chain. The Vole is a food source for mid-level predators such as snakes, foxes, hawks, etc. Its population in this region is known to cycle from around 3 to 300 individuals over a three to four year period. During population highs, this mouse can cause parts of grassland to appear mowed. Cattle grazing is not allowed to go so far as to destroy the grass canopy that covers the mice surface runways. The canopy helps protect them from predation.

Proof of the effectiveness of EBMUD nature-oriented grazing management program is seen in exclosures - fenced areas that exclude grazing in which weeds abound and the great increase in California Red-legged Frogs (*Rana aurora draytonii*), a threatened subspecies. Cattle were damaging lowland creek and ponding areas where the frogs bred. The cows were moved to upland areas, fenced off, and supplied water using a solar energy pumping system.

Many workshops have been held to train participants to conduct field studies. They have included a diversity of people—EBMUD staff, engineers, State Highway Department personnel, teachers, students, and laymen. Persons who have undergone training are given copies of the guidelines, issued map check sheets and a permit to be on watershed lands to conduct studies.

The mapping and observational procedures here described appeal strongly to the deep-seated hunting and food-gathering tradition of mankind, but instead of killing or gathering the object of the hunt, it is 'captured' by finding it and marking its place on a map. Environmental awareness and a sense of place are reinforced as the hunter attempts to relate the graphic features of the map to the physical and biological features of the actual environment where the organism was observed. Looking at the "prey" closely, and with some species handling it to determine its identity and other characteristics (as called for on the data sheets) creates an intimacy and direct personal involvement that cannot be matched by pictures or the spoken or written word. For lay persons learning about nature, and our goal of promoting nature-bonding, we have urged, for beginners, use of land marks and pacing to determine placement of locality dots on maps. This requires close attention to environmental features. Experienced and professional field persons will no doubt prefer to use Global Positioning Devices.

Some people have questioned the wisdom of involving people without professional biological training in the survey work. However, a person need not have special biological training to be a good observer. To be a good observer calls for focusing one's senses on the subject, objectivity, sometimes a bit of detective work, patience, and devotion to accuracy. Most people can meet these requirements. Through workshop training sessions interested, dedicated people can be given the guidance they need to produce results that will meet not only educational goals, but also provide useful data for biological survey studies. The great naturalist, John Muir, a keen and persistent observer of nature determined that glaciers had carved the Yosemite Valley when many professional geologists believed otherwise!

East Bay Park's Naturalist Program

By Margaret S. Kelley, Lead Supervising Naturalist

Is nature education worthwhile? Absolutely, positively and most emphatically, *YES*! And for many reasons! It gives children new experiences, broadens their backgrounds and introduces them to new and wonderful curiosities.

The founding visionaries of the East Bay Regional Park District in the San Francisco Bay area put nature education as one of three imperatives in its mission statement of 1934. The Board of Directors later set aside 700 acres of parkland for nature study to "preserve for the people of the East Bay an area that will typify for all who visit the native beauty of Northern California." General Manager, Richard E. Walpole continued, "fast disappearing in the crush of increasing population are the natural areas all Californians knew at their back door twenty and more years ago. Today's children have little opportunity to see and be close to animal and plant life displaying

APPENDIX

its charms in natural surroundings." Upon acknowledging these unfortunate changes in our culture, they proceeded to dedicate the Tilden Nature Study Area as "a sanctuary to plant and animal life. The Board of Directors also dedicates it to the people of the East Bay who once again wish to see California as it was."

Through the years, the East Bay Regional Park District has grown from its original three parks to over fifty-six. The administrators of this unique two county park system see the value of a professional naturalist staff currently stationed at six visitor centers throughout the park district's land in Alameda and Contra Costa Counties. Today the naturalists serve park visitors throughout nearly 100,000 acres of parklands.

Unique among many naturalist programs of the EBRPD is the first one that started during World War II, when one Ranger, Jack Parker, stayed in the East Bay to protect the emerging Regional Park District parklands. He manifested the vision of the EBRPD founders when in 1946 he formalized the Junior Ranger Program at the Tilden Nature Study Area. At this writing, we are in our 58th continuous year of "Mud, Sweat and Cheers!"

Junior Rangers (JRs) is focused on approximately 45 children each year ranging from ages 9 to 18 (or so). JRs meet every Saturday morning of the school year from 9 am to noon. The children apply and interview for a position in JRs each August with the program beginning in September and concluding in June. Once children are admitted into the program they usually stay until they enter college. The original one was designed for pre-teens, but there were so many JRs who did not want to leave the program, a "JR Aide" leadership position was added where teens volunteer to help teach the younger children what they learned during their tenure as a JR. The JR Aides also receive advanced training and special teen-oriented study trips. Graduation from the program results in about 10 openings for new JRs each year. Additionally, once each month JRs are invited to participate in a group backpacking overnight weekend with the naturalist. All year the JRs build their awareness of nature and wilderness skills so that they will be ready for a spring three-day backpacking study trip. Then, right after school is out in June, the JRs go on a weeklong "year-end" backpacking study trip. This immersion-type nature study and conservation education program makes a lifetime investment in quality of life as well as environmental quality.

The JR naturalists are the key to the success of this program. They facilitate opportunities to provide invaluable assistance to the EBRPD while they grow in knowledge about how nature works. They rebuild eroded trails, remove exotic weedy species of both plants and animals, set up field experiments to study animals and plant habitats, all the while having fun and learning about each other and themselves. Down to the marrow of their bones, they know that the basis for life is relationships with each other and with the natural world!

The renowned, professional naturalist staff knows that once children are out-of-doors for learning, they must conscientiously make use of activities that cannot take place in the classroom, and that allow children to participate actively in each lesson. Naturalists use methods that focus on attention, cultivating knowledge and maintaining motivation throughout children's lives. An experiential educational system of learning, also known as adventure education, is invaluable in accomplishing our goals. Experiences are jewels; in this context, the outdoor nature experiences are *diamonds* for children of all ages.

Scientific research and psychological testing have been going on for many years to determine how learning takes place. Not only were Jean Piaget and John Dewey's theories of *active* learning or *learning by doing* established as sound, but children appear to learn better through *direct experience*. The learning process is faster, what is learned is retained longer and there is greater appreciation and understanding for those things that are learned first-hand. An ancient Chinese prophet wrote, "an experience is worth a thousand pictures." Experiential learning is as old as living organisms on earth, with survival as the reward.

A *Science News* piece on September 18, 2004, cites a report by Kuo and Farber Taylor, University of Illinois, which appeared in the September 2004 *American Journal of Public Health,* confirming naturalists' intuition. The study says, "The results indicate that green environments generally improve a child's attentiveness and focus. In a separate study, Kuo and Farber Taylor sent children with ADHD on a guided, 20-minute walk along a path dominated by either natural or urban features. After the walk, children who took the nature trail performed better on a test of attention than did their counterparts who strolled in an urban setting."

Graduates from the JR program have gone on to a wide variety of professions. At the 50-year Anniversary Reunion of JRs we found many JR graduates were in professions involving nature study such as Galen Rowell, nature photographer, David Collins, Assistant General Manager for EBRPD, many Peace Corps Volunteers, AmeriCorps Volunteers, Ph.D. research biologists, wildlife biologists, naturalists, landscapers, gardeners, teachers, artists and musicians, as well as the author of this chapter.

Whether JRs go into a profession involving nature study or some other, they are staunch supporters of environmental quality, environmental education and parks. At the 50th Reunion of Junior Rangers, every person remembered and shared experiences they had when in JRs. From extended studies of spiders, insects, amphibians, birds, mammals, trails, restoration projects in watersheds to overnights and the JR song, the fun of learning in the out-of-doors inspired them to become "whole" citizens of planet earth. They thrive on their life-long joyful exploration of nature during their much appreciated escapes from the artificialities of civilization to real-life in the world of nature.

Parents often comment that, "JRs saved my child's life!" It sounds dramatic, but upon investigation, children who were struggling at home, in the neighborhood or in a structured school setting and failing, found success and acceptance in their Saturday morning outings and the weekend overnights of Junior Rangers. Parents appreciate that every JR regardless of age or gender is responsible for the same tasks: hiking, biking, studying, building, hauling, pitching a tent, setting up camp, cooking, cleaning, caring and sharing. These life skills are what it takes to have a healthy society and healthy environment. In other words, *teamwork* is essential and in the environmental sense: *relationship* is the single most important thing nature has to teach us.

APPENDIX

Orinda School District and Wagner Ranch Nature Area

By The Author with guidance from Toris Jeager, District Naturalist

The Wagner Ranch Nature Area, under the guidance of naturalist Toris Jeager, has provided the children of Orinda in the East San Francisco Bay Hills with quality nature education for over 30 years. The science programs of the Districts four Elementary Schools and a Middle School (6th through 8th grade), are closely integrated with the nature study activities of the nearby nature area.

As so often is the case, it is a dedicated and insightful person that gets things going and ensures success. Toris is such a person—a staunch defender against any changes that might mar the Ranch's long-existing and primary objective—getting children interested in nature. Like the Barstow School District Nature Program (See p. 81), several generations of students and their parents have been involved. This has led to strong community support.

Orinda is an environment where nature and humanity seem to be in sync. In many parts of its hilly landscape, homes are intermingled with free-living native and non-native vegetation. The extensive protected natural habitats of the East Bay Regional Park District and East Bay Municipal Utility District (EBMUD) are close neighbors. Indeed, the Wagner Ranch Nature Area is contiguous with wild lands of EBMUD's San Pablo Reservoir and shares a portion of the creek that supplies it. A survey reported in 1996, yielded the following counts of wildlife species found on EBMUD property—54 mammal species, 210 birds, 22 reptiles, 11 amphibians, and 684 flowering plants! The Wagner Ranch Nature Area shares a considerable number of these species. Furthermore, the nearby San Pablo Reservoir lands can be entered by permit for school group study trips. If restraint is exercised to avoid excessive development, the comfortable relationship of this community and its people and nature can persist, and can be an example for other municipalities. All should work to keep as much nature nearby as possible in our rapidly growing urban areas for the sake of this and future generations.

APPENDIX

On April 25, 2004 I attended the Wildlife Festival. The Nature Area swarmed with parents and their children, looking at garden plots of vegetables and flowers planted and cared for by the kids—including kindergartners! A large group had gathered at a display area focusing on birds, including a live disabled Great Horned Owl on a perch near ground level, affording a close look at its beautiful eyes and variegated plumage, but kept out of reach of young fingers.

I entered the nature center's greenhouse for growing and nurturing a variety of young plants, cultivated and wild, planted and tended by the children. Gathering of seeds and growing native plants is one of the activities of the nature program. The walls were decorated with kid's paintings of poppies and other flowering plants. I'm always moved by the art of children—often it seems a delightful mix of impressionism and reality.

At the festival I met Judith Adler, long supporter of the Wagner Ranch objectives. She provides nature education programs in Mount Diablo State Park, East of Wagner Ranch. These include trips that, like at Wagner Ranch, dovetail with, and make real, classroom curricula in a variety of subject areas. On the "home front", she maintains a fruit and vegetable garden/backyard wildlife habitat that is also a teaching site. To learn more about her program, see www.diablonature.org.

At the festival, I also met Bob Wisecarver, designer of brush piles he uses to attract quail and other wildlife; among them Brush Rabbits (*Sylvilagus backmani*) that increase in number thus helping to support hawks and Golden Eagles (*Aquila chrysaetos*) threatened by dwindling food supplies. Later, I visited his experimental sites at Shell Ridge Open Space near the base of Mt. Diablo to see his experiments.

Humans are not neglected. Children learn about American Indian culture, build lodges, learn about plants Indians used for food and medicine and finding them along nature paths in the area. They visit historical sites on the property and search for human artifacts. The area has been historically occupied for over 100 years and pre-historically for over five thousand.

APPENDIX

Created habitats include gardens designed to attract butterflies and hummingbirds, California native plant gardens, native grass habitats, pioneer and colonial gardens, <u>and a "frog pond."</u>

Toris contributed a few highlights concerning her many years of working with her big family of kids:

"The Wagner Ranch Nature Area provides teachers, parents, students and community with direct experiences from the past in natural and human history. The experiences also tie the past with the present in restoration studies and activities. With the formation of 'Friends of Wagner Ranch Nature Area,' we will continue all direct nature experiences into the future.

Students express it best by declaring 'I love this place. I wish I could be here every day.'

Here are some special memories: The 4th grade student who chose a five gallon Big Leaf Maple to plant near the creek, and who must have spent five one hour periods to complete the task. Who returns the next year to continue its care. He asks when leaving for middle school if I will take care of his tree. The student who thanks me for taking care of the nature area and the students, and then later writes a letter that he realizes the Nature Area takes care of both of us. The 4th grade student who, after having direct experience in growing and planting Natives, convinces his parents to landscape with natives. The Kindergarten student, when asked to paint a sign that represents Nature, paints a large red heart."

Success in environmental education may not lend itself well to formal classroom testing, but those on the front line know, first hand, the remarkable transforming effect outdoor experiences with nature can have, especially on young people. It is difficult to measure matters of the spirit.

APPENDIX

Mojave Max, the Spokes-Tortoise for the Clark County Desert Conservation Program

By Dr. Ron Marlow

In August 1989 the U. S. Fish and Wildlife Service, pursuant to the Endangered Species Act, placed the Mojave population of the desert tortoise on the list of threatened species. This action temporarily slowed development of Clark County, Nevada and its largest city Las Vegas—the fastest growing community in the United States. The community was quickly polarized into warring camps: one, consisting of private builders, rural communities and desert user groups, favoring unrestricted use of the land; the other favoring restrictions on use. Las Vegas, with a population of 700,000 in 1989 on the northeast edge of the Mojave Desert was expanding at an unprecedented pace into some of the best habitat of the tortoise. Throughout its range (in the desert of southeastern California, southern Nevada and southwestern Utah west and north of the Colorado River) desert tortoise populations had suffered dramatic population declines due to anthropogenic causes, habitat fragmentation from scattered developments, paved and dirt road proliferation, poaching, vehicle collisions, subsidized predators, and vandalism. These declines were most dramatic in the vicinity of older more established desert communities in California (Palmdale, Lancaster, Mojave, Barstow and Victorville).

In response to the listing of the tortoise Clark County undertook the development of a habitat conservation plan to obtain a permit that would allow development to continue. In 1998 Clark County received a permit for incidental "take" of tortoises and habitat to occur in some areas in exchange for conservation actions elsewhere to benefit tortoise populations. The permit allowed development to continue and required Clark County to implement a conservation plan, including a public information and education (PIE) program.

APPENDIX

Fig. 50. Mojave Max, the Mojave Max trophy, 2003 contest winner, and Clark County Commisioner, Chip Maxfield, wearing Mojave Max medals during ceremonies before the Clark County Board of Commissioners.

The mission of the PIE program is to inform the general public, interest groups and school children about the Clark County Desert Conservation Program and its message to "respect, protect and enjoy" the desert. To attract the attention of the school-age audience, the PIE committee created the cartoon character Mojave Max as the program's spokes-tortoise. Mojave Max adorned promotional materials, television public service announcements and eventually a costume and actor allowed Mojave Max to make public appearances www.mojavemax.com. It wasn't until Groundhog Day 1998 that it became clear that more than a cartoon Mojave Max was needed.

The concept of "Mojave Max: western prognosticator of spring" was originated by Las Vegas television weatherman Nathan Tannenbaum in a piece introducing Las Vegas viewers to Punxsutawney Phil, the Pennsylvania groundhog that is used to predict the coming of spring in the eastern U.S. At the end of the story,

APPENDIX

Nate, with regional pride, opined that what was needed was a western version of this story and western harbinger of spring. Tortoise biologist Jim Moore of the Nature Conservancy heard the story, called neighbor Ron Marlow (both PIE committee members) and together they conceived the Mojave Max emergence event and the role of "spokes-tortoise." Since tortoises retreat into burrows in the fall and typically do not emerge until warm days come back to the desert in late February to mid April, the emergence of Mojave Max from winter hibernation would be the event that announced the arrival of spring. If school age children could do a little research on desert seasons and tortoise biology and guess the time and day in order to win prizes, the contest could become an opportunity to learn. Thus was born the Mojave Max Emergence contest. Teaching school children about tortoise natural history, including hibernation, would be the objective of the contest; having the children guess the day and time of Mojave Max's emergence would be the means; and prizes, loot (savings bonds, t-shirts, CDs, passes to local theme parks, a class visit to the Red Rock Canyon National Conservation Area) and publicity (live television coverage of the surprise announcement in the classroom of the winner) would be the motivation.

The Bureau of Land Management volunteered to host Mojave Max at Red Rock Conservation Area Visitor Center (http://www.blm.gov/education/LearningLandscapes/menu/news04.html). A large, photogenic male tortoise that had been displaced by development was selected to be Mojave Max. The Clark County Desert Conservation Program paid for a large outdoor exhibit to house Max. The Clark County School District incorporated the contest into science curricula. Numerous local businesses, television and radio stations donated products, money and airtime to support the contest.

In the fall of 1999 Clark County Desert Conservation Program PIE coordinator, Christina Gibson, began promoting the contest through the school district internal internet system. Students were asked to fax their guesses of Mojave Max's emergence date and time. At 12:32 pm on March 15, 2000 Mojave Max emerged, and the first contest winner had guessed emergence time within a few minutes.

APPENDIX

In 2001 the Bureau of Land Management interpretive staff included Mojave Max in their classroom presentations. There was an overwhelming positive response from students. Teachers began clamoring for focused presentations from BLM, National Park Service, Tortoise Group and Clark County interpretive staffs. Cathy August, chief interpretive naturalist at BLM's Red Rock Canyon National Conservation Area received a grant from the Clark County to develop an outreach program using student interns from the University of Nevada-Las Vegas environmental education major to present desert ecology lesson featuring Mojave Max in the Clark County School District. By 2004 the outreach program was reaching more than 10,000 students each year and had won numerous national awards. In addition, Clark County, the Tortoise Group, (a local advocacy group for humane treatment of desert tortoises), and Nate Tannenbaum traveled to a dozen schools holding school-wide assemblies for another 10,000-12,000 students. The Clark County School District, the 4th largest in the U.S. serving more than 345,000, is the fastest growing school district in the nation, adding a new high school every twelve months, a middle school every 6 months and a new elementary school every 45 days. Most of these new students know nothing of the desert or the desert tortoise.

The first year only a few hundred students participated in the Mojave Max Emergence Contest, but by 2004 the number of participants had grown to several thousand. A survey conducted by Clark County to determine the effectiveness of its educational message found that more than 53% of those questioned recognized and had positive associations with Mojave Max. In 2004 the national press picked up the story of the Mojave Max Emergence Contest and students throughout the country began participating in the contest. The California Desert Managers Group, a coalition of federal and state agencies managing desert lands and resources in California adopted the Mojave Max character to promote their educational efforts. In 2005 students in southern California were able to participate in the Mojave Max emergence contest.

APPENDIX

The original mission of the PIE program was to raise public awareness among three groups: the general public, desert users and school age children. The hope was that the negative perception of the desert tortoise due to community polarization at the time of the listing would be moderated. The PIE program, especially the use of Mojave Max as the spokes-tortoise and the emergence contest, has won widespread support from school age children, more of whom recognize Mojave Max than recognize the U. S. President. One of the most gratifying results has been the enthusiasm with which the rural communities in Clark County have embraced the Max character. The rural schools are the first to request classroom visits from the BLM outreach program from Nate Tannenbaum, and from the costumed life-size Mojave Max. Several rural schools have created tortoise habitats with a resident tortoise school mascot to be used by science classes. Many rural Boy Scout troops have volunteered to help construct tortoise fencing along highways and to do desert clean-up to implement the "respect, protect and enjoy" message of the Clark County Desert Conservation Program (DCP).

The Mojave Max Emergence contest is one of the successes of the Clark County Desert Conservation Program. It represents the willingness of individuals and institutions to exploit serendipity for a higher cause. Nate Tannenbaum, a Las Vegas celebrity, created the concept and then supported it with frequent on-air plugs and an annual commitment to appear in at least a dozen school assemblies. Local biologists, educators and agency interpreters adopted the concept, devoted staff time and resources and sought extramural funding for additional programs. The community institutions and businesses embraced and supported the concept and the event. The result is that most people in Clark County now view the desert tortoise and efforts to conserve it and to "respect, protect and enjoy" the desert very positively.

APPENDIX

Dr. Dirt's "Life Lab"— An Environmental Education Program with Emphasis on Soil Food Webs

Dr. Dirt is Dr. Dale Sanders, biologist/naturalist and retired Senior Planner for U.C. Berkeley. Since retirement, he has been working in elementary schools in the Central Valley of California to foster nature education. His approach is "hands-on," having the children work directly with living systems in what he refers to as the "Life Lab"—in school gardens and simulated or natural Central Valley vernal pools. Special focus has been on soil food webs—an intricate interaction of organisms in nature, studied in both soil and in vernal pools. Because of his emphasis on the importance of soil, the children have labeled him "Dr. Dirt."

His program is an active involvement of both teachers and students in the development of topics that teach the methods of science both indoors and out. He has had the full support and assistance of teachers and administrators who recognize the value of his work. Herewith, information on some of the results of his experimental work at the Waterloo School, Linden School District, San Joaquin Co., California. Focus has been on the 5th and 6th grades. This is followed by his reflections on education and science, and the search for truth.

Understanding Ecosystems from the Ground Up

Summary of Activities and Concepts

1. "All Life."

Life is more than just plants and animals. To this end, examples were shown with microscopes and videos of bacteria, slime molds, fungi, etc. Students were told "All Life" is what you will be taught by the time you reach high school (?) or what you will be taught in college (hopefully).

APPENDIX

2. Outdoor Studies.

Outdoor studies focused on gardens prepared by the students, ponds, actual and artificial Central Valley vernal pools, and soil ecosystems.

Expected educational outcomes were understanding the nature and make up of (1) <u>food webs</u> as illustrated by both soil and vernal pool examples and (2) <u>symbiosis</u> and <u>niche</u>, two very important scientific/ecological principles. Students were to understand that life on earth has come about and been successful through cooperation (organisms working together) as well as competition between each other.

3. Gardening Studies.

Focus was on the parts of a flowering plant and how they function. As a model for flowering plant seeds, the nut of the California buckeye was used. Students germinated, propagated, and planted buckeyes on nearby Calaveras riverbed banks, and on the school grounds. The nuts came from a tree in Stockton near a busy street where the nuts were crushed and could not germinate and produce seedlings. We were on a "rescue" mission.

Students worked in teams of 5 or 6. Each team planted a spring garden in the Life Lab. They measured and mapped their beds and plantings and wrote notes in their journals of progress during the school year. They learned the difficulty in telling weeds (a plant out of place) from crops. They were soon able to sample a few radishes and carrots.

4. Vernal Pools.

Vernal pools form in spring and summer in depressions in ancient soils that have an underlying impermeable layer such as hardpan, clay, or volcanic basalt. With the onset of dry weather, the layer allows the depressions to retain water much longer than surrounding areas.

As winter rains fill the depressions, dormant freshwater invertebrates and amphibians emerge, and birds (egrets, ducks, hawks) arrive to feed on the pool's plants and animals. Vernal pools are wetlands with a complex array of specialized organisms.

In spring, as the pools dry, low-growing flowering plants produce concentric rings of beautiful changing color as different species respond to the drying conditions.

The cycle is renewed the following year as resting seeds, eggs, and cysts in the pool's mud layer emerge from dormancy. Some cysts can survive several years of drought.

In addition to studying such natural vernal pools where possible, we created artificial ones. Dried mud from a native site was placed in large plastic saucers that were activated by winter rains. Students sampled and identified insects (backswimmers, midges) and crustaceans (copepods, tadpole shrimp, etc.) that appeared in their pools. Specimens were examined, photographed, and drawn for inclusion in each student's vernal food web. It showed the place and role of each organism in the pool life system.

5. Soil Studies—The Soil Food Web.

Each student wore a graphic of several lesson plans on a T-shirt depicting Dr. Dirt's Soil Food Web, a schematic drawing of a soil ecosystem (Courtesy of the USDA, Natural Resources Conservation Service, PA1637). They filled in the answers to questions ("??") on the T-shirt as they discovered other organisms in the Life Lab. They were then able to produce a new graphic drawing with new organisms—especially any microscopic life they discovered.

6. Understanding Symbiosis and Niche.

Symbiosis and niche are important concepts in understanding ecology and the diversity of life. Symbiosis deals with interactions between all forms of life where two or more organisms live together in mutual benefit. Niche is the "job" or function of an organism within a community.

7. Scientific Journals.

Each student kept a "scientific journal" for recording measurements, drawings, notes on observations, and ideas. Each student or team wrote a story or report or did a "science" project, under the guidance of the Environmental Projects section of the San Joaquin

APPENDIX

County Office of Education's *"Jiminy Cricket's* Environmental Challenge" program. The written accounts were based on entries in their journals of scientific information, drawings, and potential ideas about what they learned and observed in the Life Lab.

8. Conducting Scientific Experiments.

Students determined how much time it took for various seeds to germinate, form root hairs, and produce stems and leaves. They grew the resulting plants until they were seedlings that could be planted in the Life Lab to discover how long it took to make flowers and produce fruit. They also devised experiments with organisms from the artificial vernal pools to better understand interactions among the species present.

A major effort throughout was to convey to students that (1) probability, logic, math, and imagination are main components to understanding science, and (2) that their understanding of the methods and purposes of sampling and the analyzing of information are part of the process involved in identifying and determining the place of organisms in the food web.

A More Detailed Explanation of Key Activities 1, 6, and 8 follows:

1. What is life?

More recent investigations into organisms living in undersea volcanic vents, "hot pots", in volcanic Yellowstone and Lassen National Parks, and other places on the earth where ecosystems are driven and fueled by other than oxygen (primarily sulfur and methane), very small, DNA-based organisms have been described. These recently recognized organisms do not fall easily into older classifications of living things. I cannot in good conscience discuss animals, plants, and bacteria as representing all life. Some of the small organisms we encountered in our soil and vernal pool studies required extending the range of our view of what is life.

Tomorrow's students need to be prepared for the "new science of life." This science reality brings into question the very science standards to which I and other teachers are supposed to teach. We explored these areas using the concepts and findings of the "Deep Green" program being developed by the University of California at Berkeley, Jepson Herbarium. We worked from their Tree of Life that subdivides life into the: Eubacteria (e.g. enterobacteria and cyanobacteria)—the Archaebacteria (e.g. halobacteria, thermoplasma, and methanobacteria); and the Eukariota (e.g. traditional plants, animals and fungi). The student's "T" shirts showed question marks at most levels of the Dr. Dirt Soil Food Web. Acting as detectives, the students discovered where many of these hidden microscopic creatures belonged in the soil ecosystem in the Life Lab.

6. Understanding Symbiosis and Ecological Niches.

The eukaryotic cell's organelles appear to have been symbionts or parasites that evolved into parts of the nucleated cell more than 3 billion years ago (mitochondria, Golgi, and endoplasmic reticulum). Life has made use of cooperative relationships, for example, since algae and fungi got together to form a new type of organisms called lichens. Without the symbiotic relationship between termites, flagellated protozoa, and bacteria, the planet would be covered miles deep in undigested cellulose and lignin—dead trees. Our objective has been that students will come to understand that they, as individuals have a niche to fill or a "job to do" in life. Finding one's place in the world is a lifelong journey that science can assist by providing a way to make the best decisions based on facts and common sense, along with an imagination and a desire to do the right thing by other creatures with which we share this planet (including humans).

8. Methods and Uses of Science.

Our students learned the distinction between guesswork and scientific proof and about levels of confidence (probability). There were discussions of how sure we need to be when presenting a hypothesis, how much confidence we need for understanding as opposed to knowledge, and ultimately, as a suggestion, how certain we need to be to use "wisdom." Life is an exercise in adaptive management—how best to conduct our lives in a virtuous fashion.

APPENDIX

Science and the Search for Truth

Who is Dr. Dirt?

My first "gig" as a volunteer science teacher was at a very small rural school in Riverbank, California. I was showing students how a scientist would use a Berlese funnel to concentrate soil organisms and then see what they look like under a microscope. The students said I should call myself Dr. Dirt—it has stuck like a piece of clay ever since.

The fact is that I have not felt particularly effective teaching college and university courses. Here was an opportunity to reach youngsters before society has imposed its idea of what science is about and how it may better serve the individual, if not society. Here was a way to influence "future scientists" to help them not lose their curiosity and inquisitiveness. To quote Albert Einstein: "Imagination is more important than knowledge." Dr. Dirt has been a volunteer science teacher for more than 9 years. Over time, the program has evolved toward a focus on the ecology of humans with an overriding question about how we should behave toward each other and other life on Earth.

How Did it Evolve?

The foregoing draft Environmental Education curriculum (the "Life Lab") was based on the concept that science can best be understood as discovery of the unknown and that methods of science can be used to discover the "facts" of a given situation. They can help us in **"a search for the truth."** A primary focus has been on the basic principles of **ecology** that can be seen through the study of "small ecosystems." As noted, the curriculum explored two small ecosystems (soil and vernal pool) in the classroom, in the student garden, and in real life locations in the nearby local landscape. It attempted to assess the new science content testing which is now included in the regular 5th grade testing under the State program. Beyond improving the student's understanding of the scientific process, my program was intended to enhance the student's sense of purpose and place. If we can impart a better feeling and understanding about humanity's niche in the Earth's ecosystems we will be successful. A

APPENDIX

major goal was to develop an ongoing effort to get students out into their local environment and to use the principles of science to better understand how their neighborhood is put together and how each student fills a niche in each ecosystem they enter and leave. Students learned to gather information and write reports. I wanted them to know that science is FUN, and anyone can become a "scientist."

The truth is that I have been evolving this viewpoint for nearly 60 years, probably starting with my "rescue" of a hatchling Western Fence Lizard from a Black Widow's web—I was hooked on **Nature!**

Philosophical Basis

The approach taken with my ideas on environmental education is guided by the concept that "understanding humanity's place in Nature leads to a virtuous way of life," from Ethics for the New Millennium by the Dalai Lama. Further, David Orr's important book focusing on environmental education, Ecological Literacy (1992), suggests, that the subject of virtue needs to become a part of what we talk about with clarity and understanding... "It will cost us something, perhaps a great deal. But, there is a far higher price waiting to be paid." Naturalist/biologist Robert C. Stebbins feels that not all is lost. In this book he has a vision of how a "nature-centered" approach to education could be a possible salvation for humanity's future.

Many philosophers/educators also share the attitude that science is probably the best example of democracy. Certainly, scientists such as: Carl Sagan, Stephen Jay Gould, and Paul Ehrlich would likely agree that "science" is not an unfathomable, underground philosophy that has no place in all of our daily lives. We are called upon almost hourly to make rational and informed decisions. Can we teach students to do this? The foregoing curriculum has been an attempt to see how we can make science and ecology less mysterious primarily by focusing on hands-on, real-life experiences in the classroom and the student garden (the Life Lab). We need to get across the idea that science is a way of viewing life, understanding life's consequences through discovery and inquiry. We should be ready to challenge our notions, ideas and beliefs. Science gives us a way for knowing, evaluating, and weighing discoverable facts. It can also give us new ideas, insights and perspectives, if we use our imaginations.

NOTES
(in three parts)

Part 1   p. 234

Part 2   p. 239

Part 3   p. 242

NOTES

Part 1

1. For about 1000 years people destroyed ants (pupae in particular) in the forests of Europe to get food for fish, birds, chicks, as well as to make a formic acid solution applied to the skin to treat rheumatism and arthritis. The ant fauna had become greatly reduced and the health of the forests had become precarious.

   On August 6, 1958, near Würzburg, Germany, I visited a forest of mixed pines, firs, and beech under study to determine what is required to maintain a healthy productive forest. The forest ant (*Formica rufopratensis*) (species name uncertain at the time) builds mound nests of forest litter, sometimes as much as 6 feet high and colonies can persist for decades. Because of the soil nitrogen content, fir trees growing near nests may be three times greater in diameter than those 15-20 feet away and have longer darker needles and longer yearly twig growth, indicating a potent effect on tree health by the ant colony.

   The ants milk aphids that feed on the forest trees. Those milked by ants are not considered harmful. The ants feed their broods with the excrement rich iron. Aphids used by ants take little from trees but give a lot to ants that in turn benefits the trees. The ants, in tunneling beneath the nest mounds, bring soil particles to the surface and bury dead ants and other organic matter, thus creating humus and enriching and aerating the soil.

   The ants also help the forest by destroying the defoliating oak tree caterpillar (*Tortrix viridana*) during outbreaks and by aiding the seed dispersal of pines.

   Since all dead trees are removed from managed forests, depriving insect controlling birds and bats of natural nesting and sheltering sites, bird and bat boxes are provided. At the time of my visit, 90 forests in Germany, involving all counties, were carrying out the ant-bat-bird program. However, also at the time, people oriented toward chemical control of forest pests tended to discount the effectiveness of the program. But chemical control destroys useful, as well as harmful insects. Those supporting the ecological approach noted that with the complex of birds, bats, ants, and good soil, one has a strong front against forest calamities over the long term.

NOTES

2. Field studies are crucial to stem the tide of declining biodiversity. There is an increasing need for dedicated and well-trained people to deal with our growing wildlife management problems, plant and animal.

Top predators have often been targeted for extermination. Fortunately, we are beginning to understand their importance in the functioning of living systems and in the maintenance of biodiversity*. With their loss and/or decline, a cascade of biological disruption, ramifying in several directions, can follow. The Gray Wolf (*Canis lupus*) was formerly wide-spread in North America. Yellowstone National Park lost the last of its wolves 80 years ago (at this writing), and they have been exterminated throughout most of the United States.

Studies in Yellowstone and surrounding areas where wolf populations have been introduced and still exist, or have been reestablished, are revealing the importance of this top predator. In its absence, elk herds increased greatly, seriously damaging, through foraging, aspen and willow groves, thus impacting needs of beaver for lodge, dam-building and food, and certain birds requiring the groves for nesting. Coyotes held in check by wolves expanded greatly, presumably with adverse effects on red fox and mid-sized predators, including wolverines, and fishers (a rare species of weasel), which are now expected to increase.

Why not have human hunters do the job of wolves? National Wildlife Federation Senior Scientist Steve Torbit answers: "People target healthy males, while top carnivores take old, young and weakened animals, as well as females. It's the females that control population growth. Even where elk are hunted, populations can continue to grow, and we are beginning to see a loss of healthy aspen grows outside of our national parks in the west. Wolves are the missing component" (See Levy, 2003).

The remarkable and fascinating story of Mark and Delia Owen's (1984) fieldwork on lions, brown hyenas, wildebeests, and other animals in the Kalahari Desert wilderness in southern Africa, is a classic example of the importance of long-term studies. Their findings and recommendations for wildlife management, if implemented, may be the best hope for the Kalahari Desert ecosystem and the future of human populations in Botswana.

---

* See Stolzenburg, 2000 (bibliography)

NOTES

3. Cause for optimism (See Sagan, 1997): There has been a sequence of gatherings of religious leaders, scientists, and legislators from many nations to try to deal with the rapidly worsening world environmental crisis.

Representatives of nearly 100 nations met at a "Global Forum of Spiritual and Parliamentary Leaders" at Oxford, April 1988 and Moscow, Jan. 1990 to address this serious issue. In Moscow, 1,300 delegates gathered in the Kremlin to hear an address by Mikhail Gorbachev on this subject. On the same day the Grand Mufti of Syria stressed the importance in Islam of "<u>birth control for the global welfare, without exploiting it at the expense of one nationality over another</u>."

The religious leaders overwhelmingly endorsed a statement titled, "Preserving and Cherishing the Earth: An Appeal for Joint Commitment in Science and Religion," drawn up by scientists in which they summarized environmental problems facing humanity. The meeting ended with a joint plan of action: "This gathering is not just an event but a step in an ongoing process in which we are irrevocably involved. So now we return home pledged to act as devoted participants in this process, nothing less than emissaries for fundamental change in attitudes and practices that have pushed our world to a perilous brink."

Since the Oxford and Moscow meetings, religious leaders in many nations have begun to act. "<u>A Joint Appeal of Science and Religion for the Environment</u>" was established, and at a meeting of scientists and leaders of the major denominations, held in New York, June 1991, a great deal of common ground was found. The religious leaders accepted ... "a prophetic responsibility" to make known the full dimensions of the environmental challenge and what is required to address it, "to the many millions we reach, teach, and council." <u>Among the steps required it was recognized that there should be "concerted efforts to slow the dramatic and dangerous growth in world population.</u>"

By 1993, the "Joint Appeal" had evolved into "The National Religious Partnership for the Environment," a coalition of faiths. Thousands of clergy and lay leaders have participated in regional training, and thousands of congregational environmental initiatives have been documented.

In Jan. 1996, a segment of the evangelical Christian community lobbied the U.S. congress in support of the Endangered Species Act (which is itself endangered). Laws protecting endangered species were described as "the Noah's Ark of our day."

Now, more than ever before, the force of the "Joint Appeal" is needed. May we hope for a great increase in its impact.

4. Spiders, despite their low appeal to many people, are especially good subjects for developing nature-bonding as long as it is understood that some species, such as the Black Widow (*Latrodectus mactans*), the Australian Red-back (*Latrodectus hasselti*), and others can be dangerous. Be careful unless you know your local species. Most are harmless to humans and widespread, have diverse and fascinating habits, are abundant, are highly important cogs in the balance of nature, and some webs are marvels of "engineering." Support strands of orb weavers have been reported to be around 50 times stronger than steel of the same diameter! Many can be observed at close range as they build their webs and feed on their prey.

Most schools and homes will have a number of species of web-builders. Two kinds of webs are common—the orb and the tray. Can species or groups be identified by their webs? On the spot, close up, photos of the resident spiders and their webs may answer this question. If the spider is out of sight, use a slender object to gently shake the web or, better, catch a small insect and place it in the web. The spider may appear (See Kaston, 1978 and Levi, 1987, for more information on spiders).

5. There is another group of insects that builds funnel-shaped sand pit-fall traps to capture ants and other small invertebrate prey. It is a group of flies (Order Diptera) of the family of Rhagionidae, the Snipe Flies, the maggot-shaped larvae of which are called wormlions or ant-tigers (genus *Vermileo*). They build their traps in fine sand, often near rocks. They occur in middle elevations in the California Sierra Nevada and in the Transverse Ranges (V. *comstocki*). Other species occur in Europe.

Ant-lions (family Myrmeleontidae) belong to an entirely different Order Neuroptera, the Nerve-winged Insects. We have here an interesting example of "convergent" evolution.

The adults of V. *comstocki* are about 6 mm in length whereas that of the common ant-lions (*Brachynemurus*) are much larger, 35-50 mm. The larva of the latter, however, is similar in length to V. *comstocki*.

NOTES

6. The Indian way of life and the buffalo have been coming back to the U.S. Great Plains as many farming families leave and corporate agriculture attempts to deal with drought and falling water tables. Efforts to reestablish, so far as possible, the natural systems are underway. Geneticist Wes Jackson, through crossing <u>deep-rooted native perennials</u> with annual crops, hopes to forge a sustainable agriculture. "Bison fed the Dakotas for centuries; annual grains—dependent on fossil fuels—won't do the same...the idea is to manage the range as if it were wild, and cattle as if they were bison."

Cowman David Lasater is in support. Tom Laseter, David's father brought a herd to eastern Colorado in 1949 and shocked his neighboring cowmen by making his spread a wildlife sanctuary where coyotes, prairie dogs, and rattlesnakes would be left unmolested. Tom said, "I like to sit back and let nature do the work...She's a hell of a lot smarter at it than we are." David carries on as a managing partner of the family ranch. "If you keep the cows moving like the bison kept moving, then the grassland has a chance to sustain its natural diversity." (See Mitchell, 2004)

NOTES

Part 2

1. Report of the State Education and Environmental Round Table. The Lieberman Hoody Report, 1998. 16486. Bernardo Center Dr., Suite 328, San Diego, Calif. 92128. Participants included: California, Colorado, Iowa, Maryland State, New Jersey, Ohio, and Pennsylvania Departments of Education; Florida Office of Environmental Education; Kentucky Environmental Education Council; Minnesota Department of Families, Children, and Learning; Minnesota GreenPrint Council; Texas Education Agency; and the Washington Office of the Superintendent of Public Instruction.

    The Executive Summary of the report focused "on a specific area of environmental education: using the <u>environment</u> as an integrating context for learning (EIC)." The group felt that this should form the foundation of environment-based education in America's schools.

    Their conclusion supports the concept of an integrating thread in education and is closely aligned with ecology as a uniting concept.

2. There is a growing transition especially in the "developed" world to a life of "mediated experience," in which technology either provides a buffer between humans and some aspect of real experience, or eliminates some annoying aspect of real-world experience all together (See Ayres, 2002). Television and computers are a persistent and pervasive source of such experience and they reach well beyond the "developed" world.

    The educational process of real-world reconnecting with nature (nature-bonding) thus faces a potent challenge. How can we instill deep respect for, and admiration of, the world of nature through direct experience when children are held mesmerized by computer games and the excitement and dazzling images and sounds portrayed on the TV screen? Some TV is helpful as found in true-to-life nature programs, but much is decidedly counter-productive because it blurs the distinction between reality and the fanciful. Parents and youth leaders must intervene. Fortunately, we have an equally potent force on our side, nature's living animals and plants in all their diversity in kind, size, color, behavior and responses that few young (or old) can resist! Field trips to our National Parks and Monuments and other outdoor natural areas with focus on wild plants and animals, can be crucial to the nature-bonding goal.

NOTES

3. Among California Indians "maximum population for a tribal area was established by the smallest amount of food available during the leanest year within the lifetime of the members of the tribe at anytime." They had devised and practiced various methods of population control to keep their numbers within the limits of the food that their habitat could supply. These included abortion, the drinking of plant infusions to prevent conception, and infanticide (See Heizer and Elsasser, 1980, p.26).

4. The Promethean Myth (Cowles, 1977). "Surprising insights can often be gleaned from even the most primitive myths. In a discussion of early Greek mythology, George Boas and A. O. Lovejoy (1935) reflected on mankind's most ancient views regarding its status in the world:

> The elements in Greek mythology which are most significant are two—the legends of the Golden Age and of the ages that came after it, and the story of Prometheus, a culture hero who brought the benefits of agriculture and fire to his people.
>
> Similar myths are found in almost every recorded culture, past and present, and thus since history began most men have divided themselves into two camps—the optimists and the pessimists. The optimists have always believed in a never-ending progress toward the day when man will have mastered "the arts of life" completely; the pessimists have always harked back to a "Golden Age" when life was better, when the streams ran clean and pure, the world was clothed in verdure, gardens yielded luxuriant crops, and herds waxed fat on the hillsides.
>
> What were the sources of these still-prevalent Promethean myths, in all their variations, and of the opposite, the myth of the "Golden Age?" It appears probable that neither is wholly mythical. Both represent the distillation of actual historical experiences in a past that was replete with alternating periods of hardship and comfort, of chaos and ordered stability. From such historical sources as the Bible and Egyptian records, plus observations of events well into the twentieth century, such as the Russian famine of the early 1930's and that of Bengal in 1943, we know that periods of famine have almost always followed times of abundance; and we can find no reason to believe that they will not continue to do so. The food and energy crises of 1973-74 illustrate the inevitability of these resource trends and rhythms.

Today, the optimistic Procrusteans hold center-stage, their faith and admiration turned to the new tribal-culture heroes- the big-business tycoon, the inventor, the scientist, the technologist, the manufacturer. The optimists accept the implicit promises of their new gods to provide not only magical new gadgets, but also limitless new substitutes for food and for diminishing resources. And they seem to imagine that all succeeding generations will enjoy still more glorious products of Promethean invention. The faith of these optimists, however, is a fantasy - a delusion perpetuated by seeing only the promise of science and refusing to recognize its warnings, among them the admonition that since men themselves are biological entities ultimately dependent on a biological environment for their own ultimate salvation, they must exist in harmony with it and live in equity with it."

5. The Mighty Engine With No Brakes: In the 20th century, modern medicine and public health programs lowered mortality, and modern agriculture fed the rising numbers, but too little was done, too late, to lower fertility. That created a fundamental demographic imbalance. The resulting population growth has dwarfed all previous human experience. World population quadrupled in one century, a change so astonishing that it has altered—or should have altered—our assumptions as to the human connection with the rest of the planet. Are we plunging toward a collapse because of that very success? Philosophers since John Stewart Mill have warned against the illusion of perpetual growth. Endlessly growing numbers cannot enjoy endlessly growing consumption. There is a mathematical platitude that post-Keynesian economists ignore: material growth at some point becomes a logical impossibility on a finite planet. When? (From Grant, 2004. The New American Century.)

NOTES

Part 3

1. **Growthmania: The Durable Bubble**

"Before the 1300s, the idea of unlimited growth hardly figured in human thinking. Then came the Renaissance, which led to the Age of Exploration to the new world and new wealth. It started the agricultural and industrial revolutions and set in motion a worldwide scientific enterprise that is still accelerating. It began a period of enrichment and growth without parallel in human history.

The period has lasted, with minor interruptions, for six centuries. That success led gradually to a widespread conviction that growth is the natural and desirable order of things, and forever benign. Enter the Romantic Era and its sense of limitless horizons, and the "Age of Exuberance" (to borrow William Catton's term). Western civilization is still in that mode and is teaching it to the East and South.

It is a formidable belief system, but its proponents have forgotten that its origins were not in population growth, but in the Black Death, the most widespread and severe population collapse in human history. *Brutally, it readjusted the ratio of people to land.* (Italics mine.) The surviving peasants found themselves with more farmland and more wealth. New wealth flowed into the depopulated cities. The institutional constraints of Feudalism were swept away and replaced by the system now identified with capitalism. The subsequent Age of exploration further improved the ratio of land to people by opening access to the new world, which has more than quadrupled the arable land available to Europeans." (From Grant, 2004)

"The principles of a more rational economics were laid out over 30 years ago by Herman Daly, who has been an economics professor in various U.S. universities as well as a senior adviser in the World Bank. Daly argued for a three-stage process. First, he said, we need to ensure that the total scale of human activity is ecologically sustainable. Second, we should distribute resources and property rights fairly, both within this generation and between generations, also taking into account the needs of other species. Finally, Daly proposed, we should try to allocate resources as efficiently as possible within these constraints. So there is a role for markets in ensuring efficient allocation of resources, but first, science must determine the scale of resource allocation we can responsibly allow and society needs to work out the principles of fairness within which markets can operate. This is a much more sophisticated approach than taken by most economists, who Daly believes 'chase an assumption of infinite wants along the road of infinite growth.' If we know that the present level of resources use is straining the capacity of natural systems, encouraging further use of these resources is totally irresponsible." (From Lowe, 2005, p.35 of his book.)

2. Comments on the trial course in Conservation Education at U.C. Berkeley given by Professor Lawrence Lowery, U.C Berkeley School of Education, and me, spring and fall 1968, by one of our students, Jennifer Meax (now J. M. White). June 1968 (See also p.71).

"Conservation education has been one of the most relevant courses that is being offered at this university, and I feel very lucky that I could take some part in it. This was one of the most profitable courses I have taken in my college years. Here is a course that is directly and enthusiastically (despite setbacks and difficult to manage children) trying to tackle the most important area of education. I only hope that we are not too late.

I think the discussion we had last week sums up pretty well what I think the direction of the course should be. It should be directed at developing techniques and experiences and experiments which will expose elementary students to the natural world—stressing the local environments, and more particularly man's activity in that environment. I favor an applied-ecology approach, which uses as its core two present concepts—man's management and mismanagement of his surroundings. For if we are to try to salvage and save the natural world from man's slaughter we must begin yesterday and a theoretical approach will not convey the immediacy, enormity or the reality of ecological problems which face us today. I strongly feel that the most important thing that can be done in the elementary grades is to help children understand man as a part of the universe. Man does not stand above the natural systems but is an important part of them. Important largely in the fact that he can destroy the delicate balance of nature more easily and more thoroughly than any other animal.

The course was discontinued after two summer sessions because its appropriateness as an offering by MVZ was in question. Perhaps now the value of a collaborative effort between departments of science and education in developing science topics in outdoor biology for elementary school children would more likely to be supported.

NOTES

3. On flights (2005) to Tucson and Palm Springs from the San Francisco Bay area, after a gap of several decades, I was appalled at the tremendous increase in ground coverage by housing developments that have occurred in that brief period. I also noted around some of the subdivisions a lacework of ORV tracks that were enveloping the small hills and patches of nature left nearby.

With a ranger-led group of members of the Tortoise Council (an agency working to protect the Desert Tortoise), we stood in an upland area of the Saguaro National Monument in Tucson discussing the plight of the Tortoise—now isolated in mountain areas because surrounding developments have cut off its movements between ranges (Other species such as Bighorn Sheep and Mountain Lions have also been affected). Such isolation will ultimately result in inbreeding and species decline.

Our group felt at least a partial solution to the problem is accelerating the concept of corridors that connect protected areas—now, however, too late for some. A wise future course in environmental planning would seek to greatly increase networks of corridors planet-wide between nature reserves to allow terrestrial and perhaps also some winged species (as needed) to maintain gene flow between populations. Such corridors would need to be closed to any human activities that might interfere with their primary function.

As our meeting drew to a close the possibility of transporting tortoise and other isolated species from one mountain range to another was considered but this would be a chancy stop gap measure. Again our numbers and demands came to mind and the overriding requirement that we <u>must</u> reduce our numbers.

Bibliography

Ableman, M. 2000. The quiet revolution. Urban agriculture: taking roots in cities. Earth Island Journal, Vol.15, No.3.

Acorn Naturalists. 2004. Resources for the trail and classroom. 155 El Camino Real, Tustin, CA 92780. An excellent catalogue for guidebooks, teaching materials, etc. for those interested in science and nature.

Anderson, M. 2005. Tending the wild. Native American knowledge and the management of California's natural resources. U.C. Press.

Ausubel, K. and J. Harpignies. 2004. Nature's operating instructions. See Part III Graffiti in the book of life: Geneting engenering and the vandalism of nature.

Ayres, E. 2002. Out of touch. World Watch. Working for a sustainable future. Vol.15, No.5, p.3.

Bainbridge, D. and R Virginia. Desert soils and soil biota, in Latting, J. and P.G.

Rowlands, eds., 1998, The California Desert, an introduction to natural resources and man's impact. California Native Plant Society.

Barbour, I. (ed.) 1973. Western man and environmental ethics. Attitudes toward nature and technology. Addison-Wesley Publishing Co.

Bartholomew, G. 1986. The role of natural history in contemporary biology.

Bioscience, Vol. 38, No. 5, pp. 324-329.

Benz, R. 2000. Ecology and evolution: Island of change. National Science Teachers , Association Press.

Berger, C. 2003. When gardeners grow wild. National Wildlife. Vol. 41, No.4.

Berry, W. 2004. Citizens dissent: Security, morality, and leadership in an age of terror. Thoughts in presence of fear: Three essays for a changed world.

BIBLIOGRAPHY

Biological Science Curriculum Study. BSCS Green Version. 2002. Biological science: An ecological approach. Eighth Edition. Kendall/Hunt Publishing Col, 2460 Kerper Blvd. P.O. Box 539. Dubuque, Iowa 52001.

Blackburn, T. and K. Anderson 1993. Before the wilderness. Environmental management by native Californians. A Ballena Press Publication, 823 Valparaiso Ave., Menlo Park, CA. 94025.

Boas, G. and A. Lovejoy. 1935. Primitive and related ideas in antiquity. Baltimore, Johns Hopkins Press.

Borror, D. and R. White. 1998. A field guide to the insects. Houghton Mifflin Co.

Bowen, J. 1900. English words as spoken and written. New York, Globe School Book Co.

Brown, L. 1995. Who will feed China? World-Watch, October 1994.

– 1998. State of the world. 170.

– 2006. Plan B. 2.0 Updated and expanded. Rescuing a planet under stress and a civilization in trouble. Earth Policy Institute. W. W. Norton & Co.

Brown, L., et al. 1984 and continuing. State of the World. A World Watch Institute report on progress toward a sustainable society. New York, W.W. Norton & Company, Inc. Vol.1, No.2.

California Journal of Science Education: Controversy in the classroom II: Evolution. Vol. 1, Issue 2—Spring 2001. Teachers Association, 3550 Watt Ave., Ste. 120, Sacramento, CA 95821-2666.

Carson, R. 1998. The sense of wonder. Harper Collins, reprint edition.

Carter, J. 2005. Our endangered values. America's moral crisis. Simon and Schuster.

Cincotta, R., R. Engelman, and D. Anastasion. 2003. The security demographic. Population and civil conflict after the cold war. Erwin Population Action International.

Clarkson, K. 2006. Act locally, make friends globally. How a salamander brought Russian and American students together. Tideline. San Francisco Bay National Wildlife Complex. Vol. 26, No. 1.

Comstock, A., V. Rockcastle. 1986. Handbook of nature-study. For teachers and Parents. Ithaca, N.Y., Comstock Publishing Company.

BIBLIOGRAPHY

Cowles, R. 1977. Desert Journal, Epilogue: The Promethian Myth, Part II, 23, p.248-255, The Regents of teh University of California Press.

Daly, H. 1977. Steady state economics: the economics of biophysical equilibrium and moral growth. San Francisco, W.H. Freeman & Co.

Darlington, D. 2003. Untracked Utah. Off-road vehicle sales are booming, but wilderness isn't a drive through experience. Sierra Jan/Feb. 2003.

Diamond, Jared, 2005. Collapse. How societies choose to fail or succeed. Viking. Penguin Group.

Diamond, Judy ed., 2006. Virus and the whale. Exploring evolution in creatures small and large. National Science Teachers Association Press.

Dobzhansky, T. 1973. Nothing in biology makes sense except in the light of evolution. Amer. Bio. Teacher, 35: 125-129.

Earthwise. Fall 2005. An edible legacy. Union of Concerned Scientists, Vol. 7, No. 4.

Edwards, B. 1989. Drawing on the right side of the brain. Jeremy P. Tarcher/Putnam Penguin Putman Inc. New York.

Ehrlich, P. 1968. The population bomb. Sierra Club, Ballantine — 1985. Human ecology for introductory biology courses: An overview

– American Zoologist 25: 379-394. Suggestions on what topics should be emphasized in covering the population-resource-environment dilemma and how they might be introduced to students.

– 1987. The machinery of nature. The living world around us--and how it works. New York, Simon and Schuster. Provides an easily understood background to principles of ecology and how they operate in nature and constrain the activities of humanity. An ideal source book for the nature-centered program espoused in this essay.

Ehrlich, P. and A. Ehrlich. 1990. The population explosion. Simon and Schuster, New York, NY.

– 2004. One with Nineveh. Island Press, Washington, D.C.

– 2008. The dominant animal. Island Press.

Elson, R. 1964. Guardians of tradition. American school books of the nineteenth century. Lincoln, University of Nebraska Press.

Engelman, R. 2008. More: Population, nature, and what women want. Island Press, 2008. See unnatural increase? A short history of population trends and influences.

BIBLIOGRAPHY

Faden, M. 2004. The deserts living crust. The Survivor, Spring 2004. P.O. Box 20991, Oakland, CA. 24620-0991.

Flannery, T. 1994. The future eaters. Reed Books, Chatswood.

Gardner, G. 2002. Invoking the spirit. Religion and spirituality in the quest for a sustainable world. World Watch Paper 164.

– 2006. Inspiring progress. Religion's contributions to sustainable development. A Worldwatch Book. W. W. Norton & Co.

Gasana, J. 2002. Remember Rwanda? World-Watch, Vol.15, No.5.

Gendler, L. 2006. Transgenic crops. Berkeley Scientific, Vol. 10, Issue 1, Spring 2006.

Geyer, G. 1996. Americans no more - The death of citizenship. New York, NY, U.S.A. Grove/Atlantic, Incorporated, 1996

Global Senarios Group. 2002. Great transition. Stockholm Environment Institute, Boston.

Gorman, P. 2003. Awakenings. Spiritual perspectives on conservation. Nature Conservancy, Vol. 53, No. 1, p.22

Grant, B. and L. Wiseman. 2002. Decline in melanism in American Peppered Moths following clean air legislation. Journal of Heredity 93.

Grant, L. 2000. Too many people. The case for reversing growth. The book is dedicated to the two-child family. Seven Locks Press, P.O. Box 25689, Santa Ana, CA. 92799.

– 2004. The new American century and the end of fossil fuels, Part2. Twilight or dawn? November, NPG Forum. Negative Population Growth, 2861 Duke St., Suite 36, Alexandria, VA. 22314.

– 2005. Social security and the fear of aging. August, NPG Forum, Negative population growth and NPG Booknote The Wrong Apocalypse—Grant's review of P.G. Peterson's "Gray Dawn", Random House, 1999 on population aging.

– 2007. Valedictory: The age of overshoot. Negative Population Growth. 2861 Duke St., Suite 36, Alexandria, VA. 22314.

Grinnell, J. 1943. Joseph Grinnell's philosophy of nature. University of California Press.

Grobman, A. and H. Grobman. 1989. A battle for people's minds: creationism and evolution. The American Biology Teacher, Vol. 51, No. 6.

BIBLIOGRAPHY

Grossi, R. 2003. American farmland trust, 1200 18th street. NW Suite 800, Washington D.C. 20036. Letter section July 16, 2003.

Gould, S. 1999. Dorothy, its really Oz. Time, Aug. 23, 1999, p.59.

Hatch, W. 2008, A frogs life: Seen and heard. Life cycle of the Pacific Treefrog (*Pseudacris regilla*). DVD by Warren A. Hatch, Box 9224, Portland, OR 97207-9224

Halweil, B. 2002. Home grown. The case for local food in a global market. Worldwatch Paper 163.

– 2004. Eat here. Reclaiming homegrown pleasures in a global market. World Watch. W. W. Norton & Company, New York and London.

– 2005. The irony of climate. Archeologists suspect that a shift in the planet's climate thousands of years ago gave birth to agriculture. How climate change could spell the end of farming as we know it. World Watch March/April 2005.

Hancock, J. 1991. Biology is outdoors! A comprehensive resource for studying school environments. J. Weston Walch Publishers. Contains lists of field guids and other nature references.

Hardin, G. 1968. The tragedy of the commons. Science, Vol. 162, No. 3859, pp. 1243-1248.

Heizer, R. and A. Elsasser 1980. The natural world of the California Indians. University of California Press.

Henley, T. 1996. Rediscovery: Ancient pathways—new directions. A guidebook to outdoor education. Western Canada Wilderness Committee, 20 Water St., Vancouver, British Columbia, Canada V6B1A4.

Hickman, J., ed. 1993. The Jepson manual. Higher plants of California. University of California Press.

Hubbard, F. 1955. Animal friends of the Sierra. Awani Press, Yosemite National Park.

Jones, Van. 2008. The green collar economy: How one solution can fix our two biggest problems. Harper Collins Publisher, New York, NY 10022.

Kahn Jr., P. and S. Kellert (eds.) 2002. Children and nature: psychological, sociocultural, and evolutionary investigations. MIT Press. Children pay a price for ignorance about their place in the world of nature.

Kaston, B. 1978. How to know the spiders. McGraw-Hill Science/Engineering/Math; 3rd edition.

BIBLIOGRAPHY

Kellert, S. and E. Wilson (eds). 1995. The biophilia hypothesis. Island Press, Washington, DC.

Kennedy, R. 2005. Crimes against nature.

Kettlewell, H. 1958. Vol. 12, Pt. 1 Scientific American 1959. Heredity Vol. 200, No. 3

Lader, L. 1971. Breeding ourselves to death—special edition printing 2002 Seven Locks Press, P.O. Box 25689, Santa Ana, CA 92799. NPG 30 Anniversary Special Edition—Contact Negative Population Growth. 1717 Massachusetts Ave., NW Suite 101, Wash., D.C. 20036.

Latting, J. and P. Rowlands eds. 1995. The California Desert: An introduction to natural resources and man's impact. June Latting Books. 320 Margvilla Dr., Riverside Press, Riverside CA.

Lawson, L. 2005. City bountiful. A century of community gardening in America. U.C. Press.

Lehmer, A. and J. Phelan 2003. World sustainability hearing special report: Grassroots globalization network. Earth Island Journal, 2003, Vol.18, No.1, p.23.

Leopold, A. 1970. A Sand County almanac and sketches here and there. New York: Oxford University Press.

Leslie, C. 1995. Nature drawing. Kendall/Hunt Publishing.

Levi, H. and L. 1987. Spiders and their kin 2nd ed. Alfred A. Knopf.

Levy, S. 2003. A top dog takes over. Exterminated from Yellowstone National Park eight decades ago, gray wolves are back—and boosting the park's biodiversity. National Wildlife. National Wildlife Federation, Oct./Nov. 2003. 11100 Wildlife Center Drive, Reston, VA 20190-5362.

– 2005. Fires down under. Onearth, Vol.26,No.4

Lieberman, G. and L. Hoody. 1998. Closing the achievement gap. Using the environment as an integrating context for learning. Science Wizards, 13648 Jackrabbit Road, Poway, CA. 92064.

Lossing, B. 1866. A primary history of the United States. New York: Mason Bros.

Louv, R. 2005. Last child in the woods: Saving our children from nature deficit disorder. Algonquin Books of Chapel Hill. The world is not entirely available from a keyboard.

– 2006. Leave no child inside. The remedy for environmental despair is as close as the front door. Sierra July/August 2006.

Lovelock, J. 1972. Gaia as seen through the atmosphere. Atmos. Environ. 6. 1979. Gaia. A new look at life on earth. Oxford University Press.

– 1988. In E. O. Wilson (ed.). 1988. Biodiversity. National Academy Press, Washington D.C.

Lowe, I. 2005. A big fix. Radical solutions for Australia's environmental crisis. The Public Interest Series. Black Inc. Gives full recognition to the problem of human numbers.

Malthus, T. 1798. An essay on the principle of population. 1992. Cambridge University Press. (Re-print).

Mann, D. 2002. The United Nations World Summit Development— A counter-productive exercise in futility. Negative Population Growth, Inc. 1608 20th St. NW, Suite 200, Washington, D.C. 20009. Overpopulation issue ignored.

Martin, P. 2005. Twilight of the mammoths. Ice age extinctions and the rewilding of America. U.C. Press.

Mathews, J. (ed.) 1990. Preserving the global environment: The challenge of shared leadership. W. W. Norton & Co.

McChesney, R. 1999. Rich media, poor democracy: Communication politics in dubious times. New Press.

McCluskey, M. 2004. Case dismissed: land pays the price. Earth Island Journal, Vol. 19, No. 3.

McManus, R. 2004. I call it a delicious revolution. Sierra, Nov./Dec.

Meadows, D., D. Meadows, and J. Randers. 1992. Beyond the limits. Confronting global collapse. Envisioning a sustainable future. Chelsea Green Publishing Co., P.O. Box 130, Post Mills, VT. 05058.

Meadows, D., J. Randers, W. Behrens III. 2004. The limits to growth: The 30-year update. A report for the club of Rome's project on the predicament of manking. Chelsea Green Publishing Company. White River Junction, VT. See also a synopsis to Limits to Growth (the 30 years update) available by contacting Diana White at the Sustainability Institute, 3 Linden Road, Hartland, Vermont, 05048. Website http://sustainer.org/limits.

BIBLIOGRAPHY

Milne, L. and L. Milne. 1980. Field guide to North American insects and spiders (Audubon Society Field Guide). Alfred A. Knopf.

Millar, H. 2008. Restoring rare beauties. National Wildlife Federation. June/July, Vol. 46 (4)

Mitchell, J. 2004. A profound change is in the wind: the unsettling of the Great Plains. National Geographic. May, 2004.

Moore, J. 1985. Science as a way of knowing: Human ecology. American Zoologist 25(2): 377-378.

Murie, O. 1976. Animal tracks. 2nd Ed. Peterson Field Guide Series. Houghton Mifflin Co., 215 Park Ave., New York, New York. 10003.

Myers, N. and J. Kent. 2005. The new atlas of planet management. (Revised edition). U.C. Press and Gaia Books.

Nash, R. 1982. Wilderness and the American mind. New Haven, Yale University Press,. (3rd ed.)

National Academy of Sciences 1998. Teaching about evolution and the nature of science.

National Academy Press, 2101 Constitution Ave., NW, Box 285, Washington D.C. 20055.

Nowak, M. 1998. Our demographic future: why population policy matters to America. Negative Population Growth, Inc., 1608 20th St., NW, Suite 200, Wash., D.C. 20009.

Orr, D. 1992. Ecological literacy. Education and the transition to the postmodern world. State University of New York Press, Albany.

– 2002. Four challenges of sustainability. Conservation Biology, Vol. 16, No. 6, pp. 1457-1460.

– 2003. The constitution of nature. Conservation Biology, Vol. 17, No. 6.

Owens, M. and D. Owens. 1984. Cry of the Kalahari. Houghton Mifflin Co., 215 Park Ave. South, New York, NY.

Pfeiffer, D. 2003. Eating fossil fuels. Carrying Capacity Network.

Pianka, E. and L. Vitt. 2003. Lizards: Windows to the evolution of diversity. The Regents of the University of California.

BIBLIOGRAPHY

Population Crisis. 1965-1968. Hearings before the subcommittee on foreign expenditures of the committee on government operations United States Senate—Eighty-ninth and Ninetieth Congress. A bill to reorganize the Department of State and the Department of Health, Education, and Welfare. S.1676. 16 booklets were published, with "Population Crisis" in bold face on the spine of each.

Richardson, J. 1813. The American reader. 2nd ed. Boston, Lincoln and Edmands.

Ridgley, H. 2005. When "weed laws" make homeowners outlaws. National Wildlife, December/January.

Royte, E. 2003. Don't spoil the soil. On Earth. National Resource Defense Council. Vol.23, No.1, p.26.

Sagan, C. 1996. The demon-haunted world. Science as a candle in the dark. Ballantine Books, New York. 1997. Billions and billions. The Ballantine Publishing Group.

Sanders, R. 2001. Growing science lessons. Cal Neighbors. A news letter for neighbors of the University of California, Berkeley. Fall 2001.

Shand, H. and K. Wetter 2006. Shrinking science: An introduction to nanotechnology.

State of the World. Worldwatch Institute. W. W. Norton & Co., New York and London.

– 2006-2007. Nanotechnology takes off. Vital Signs Worldwatch Institute.

Simpson, G. 1960. The world into which Darwin led us. Science, Vol. 131, No. 3405, pp. 966-974.

Skerham, J. and C. Nelson 2000. The creation controversy and the science classroom. National Science Teachers Association Press.

Snell, M. 2003. Life study. How nature nurtures students at an inner-city high school. Growing up green at Balboa High. Sierra, Nov./Dec.

– 2004. A fine balance. Cultivating smaller families and healthier farms in Ecuador's highlands. Sierra, Jan./Feb.

Stebbins, C. 1913-1924. The principles of agriculture through the school and the home garden. New York, The Macmillan Co.

BIBLIOGRAPHY

Stebbins, R. 1949. Speciation in salamanders of the plethodontid genus *Ensatina*. University of California Press, Berkeley, California.

– 1954. Natural history of the salamanders of the plethodontid genus *Ensatina*. University of California Press, Berkeley, California.

Stebbins, R. 1971. The loss of biological diversity. In Brown, M. ed., The social responsibility of the scientist. The Free Press, New York and Collier Macmillan, London.

– 1974. Off-road vehicles and the fragile desert. The Amer. Biol. Teacher, Vol. 36, No. 4, pp. 203-208 and No. 5 pp. 294-304.

– 2003. Western reptiles and amphibians 3rd ed. Peterson Field Guide Series. Houghton Mifflin Co., 215 Park Avenue, New York, New York 10003.

Stebbins, R. and B. Allen 1975. Simulating evolution. The Amer. Biol. Teacher, Vol. 37, No. 4, pp. 206-211.

Stebbins, R. and N. Cohen. 1976. Off-road menace. A survey of damage in California. Sierra Club Bulletin, July/Aug.

– 1995. A natural history of amphibians. Princeton University Press. Princeton, New Jersey.

Stebbins, R. and EBMUD staff 1996. Biological survey studies for the East Bay Municipal Utility District. Guidelines I and II. Published by the District.

Stebbins, R., T. Papenfuss, and F. Amamoto 1978. Teaching and research in the California Desert. Institute of Governmental Studies, Univ. of Calif., Berkeley, California.

Stokes, D. and L. Stokes. 1985. A guide to enjoying wildlife. Little, Brown and Company. Boston, Toronto.

Stolzenburg, W. 2002. Playing God in a tide pool. A simple little extinction experiment with big consequences. Nature Conservancy, Vol. 54, No. 4.

Stone, M. and Z. Barlow eds. 2005. Ecological literacy. Educating our children for a sustainable world. Sierra Club Books. Many contributors.

Stuart, J. and J. Sawyer. 2001. Trees and shrubs of California. California natural history guides. University of California Press.

Suzuki, D. 1994. Time to change. Stoddard publishing Co. Limited, 34 Lesmill Rd. M3B2T6. Toronto, Canada.

BIBLIOGRAPHY

Taylor, J. 1963. Effect of field trips and an ecological approach on science attitudes and knowledge. MA in education, University of California, Berkeley, Education Library.

The Reporter. 2004. Why anxiety over low fertility? Vol.36, No.3. A production of Population Connection. (Formerly the ZPG Reporter) 1400 Sixteenth Street, N.W. Suite 320, Washington, DC 20036.

– 2008. Growing needs, shrinking fields. Vol. 40, Issue 3

Toland, J. 1976. Adolf Hitler. Vols I and II. Doubleday and Co., Inc., Garden City, New York.

Tooland, D. 2001. At home in the cosmos. Orbis Books, P.O. Box 308., Maryknoll, New York, 10545-0308.

Webb, R. H. and H. G. Wilshire (eds.). 1983. Environmental effects of off-road vehicles. Springer-Verlag, New York.

Wellington, Department of Education. 1967 Biology. Science for Forms I and II Teacher's Guide. Food Chains, p. 12.

Wheat, F. 1999. California desert miracle. The fight for desert parks and wilderness. Sunbelt Publications, San Diego, California.

White, L., Jr. 1967. The historical roots of our ecologic crisis. Science, Vol. 155, No. 3767. American Association for the Advancement of Science.

– 1973. "Continuing the conversation," in Western man and environmental ethics: Attitudes toward nature and technology, ed. Ian G. Barbour. Reading, Mass: Addison-Westey, 61.

Wilson, E.O. 1984. Biophilia: The human bond with other species. Harvard Univ. Press. Cambridge, Mass.

– 2002. The future of life. Alfred A. Knopt.

– 2006. The Creation. An appeal to save life on earth. Norton.

Wilson, E.O. and F.M. Peter (eds.) 1988. Biodiversity. xiii + 521pp. National Academy Press, 2102 Constitution Ave., NW., Washington, DC 20418.

World-Watch. 2004. Population and its discontents. Vol.17, No.5.

– 2008. Population forum: From Malthus to birth bribes. The role of women's empowerment. U.S. public attitudes. A Kenya urbanization link. Security implications. Vision for a sustainable world. Vol. 21, No. 5

– 2008. Article by John Bermingham and Raymond Reddy. Vol. 21, No. 6.